Sudarno

Nitrification in Fixed Bed Reactors Treating Saline Wastewater

Karlsruher Berichte zur Ingenieurbiologie
Band 47

Institut für Ingenieurbiologie und Biotechnologie des Abwassers
Karlsruher Institut für Technologie

Herausgeber: Prof. Dr. rer. nat. J. Winter

Nitrification in Fixed Bed Reactors Treating Saline Wastewater

by
Sudarno

KIT Scientific Publishing

Dissertation, genehmigt von der
Fakultät für Bauingenieur-, Geo- und Umweltwissenschaften
des Karlsruher Instituts für Technologie, 2011
Referenten: Prof. Dr. rer. nat. J. Winter
Prof. Dr.-Ing. E. h. H. H. Hahn, PhD

Impressum

Karlsruher Institut für Technologie (KIT)
KIT Scientific Publishing
Straße am Forum 2
D-76131 Karlsruhe
www.ksp.kit.edu

KIT – Universität des Landes Baden-Württemberg und nationales
Forschungszentrum in der Helmholtz-Gemeinschaft

KIT Scientific Publishing 2011
Print on Demand

Gedruckt mit Unterstützung
des Deutschen Akademischen Austauschdienstes

ISSN: 0172-8709
ISBN: 978-3-86644-691-5

Nitrification in Fixed Bed Reactors Treating Saline Wastewater

Zur Erlangung des akademischen Grades eines

DOKTOR-INGENIEURS

Von der Fakultät für

Bauingenieur-, Geo- und Umweltwissenschaften,

Karlsruher Institut für Technologie (KIT)

genehmigte

DISSERTATION

von

Sudarno, MSc

Boyolali - Indonesien

Tag der mündlichen Prüfung : 6. May 2011

Hauptreferent : Prof. Dr. rer.nat. J. Winter

Korreferent : Prof. Dr.-Ing. E.h. H. H. Hahn, PhD

Karlsruhe 2011

Acknowledgments

First of all I would like to thank Deutschen Akademischen Austauschdienstes (DAAD) which sponsored more than four years of my doctoral in Karlsruhe Institute of Technology (KIT), also Max Buchner Research Foundation-DECHEMA for approving a grant to perform the laboratory work.

I would like to express my sincere gratitude to my promotor Prof. Dr. rer. nat. Josef Winter for giving me the opportunities to study and finish my doctoral. I am grateful to em. Prof. Dr.-Ing E.h. H. H. Hahn, PhD, for being my co-promotor and writing recommendation letters to DAAD for my scholarship extensions. Special thanks go PD Dr. rer. nat. Claudia Gallert for her valuable discussions, her inspring and motivating guidance throughout this research also for hospitality to my family.

My thanks to my former and present colleagues at the Institut für Ingenieurbiologie und Biotechnologie des Abwassers: Dr. rer. nat. Mini Bajaj, Daniel Jost, Renate Anschütz, Dominik Freikowski, Rita Seith, Dr. rer. nat. Stephan Bathe, Monika Felchner-Zwirello, Dr. rer. nat. Susan Schmidt, Dr. Ing. Satoto Nayono, Chaoran Li, Daniel Bonefas and Andreas Morio for the help in the academic and laboratory activities.

I would also like to thank my Indonesian friends here in Karlsruhe, especially Pak Nurdin, Mbak Yuli for *ihre Hilfe für mich ohne Wenn und Aber*, also Dr.-Ing. Muhammad Ikhwan and family. Many thanks also to Dr Petra Roth from International Office KIT – Karlsruhe and Dr. Med. Armin Hettich.

I also want to ackowledge to all DAAD scholarship holders aus Indonesien for academic year 2006/2007 especially Dr. rer. nat Endang Lukitaningsih, Dr. Ratna C. Purwestri, Dr. Yudi Sukmayadi, Dr. Zaenal Burhan and Dr. rer. nat. Telly Savalas. Furthermore, my thanks to my colleagues at the Department of Environmetal Engineering – University of Diponegoro – Semarang.

All my life I will be deepy indebted to my parents Mbah Nenek and Mbah Kakung, my family, my parent in law Opung Besar and Opung Kecil and my family in law, for their love, emotional support and care.

My greatest thanks are to my wife, Maya Lubis, and to our lovely sohn Risang Wisesa I do not know, how to thank you for your love and support. *Wir schaffen es zusammen, denn nur zusammen sind wir stark und können die schwierigsten Zeiten überstehen.*

Abstract

Two fixed bed reactors for nitrification with either polyethylene/clay sinter lamellas (FBR A) or porous ceramic rings (FBR B) were continously run for treating synthetic saline wastewater. Seawater from Hafen Büsum was used as an inoculum for the both reactors. The performance of the system was evaluated under different operating conditions such as ammonia loading rates (ALRs), ± pH adjustment, and at different upflow velocities.

No physical damage of supporting material due to biochemical or physical processes in the FBRs was found during the whole experiment. A better overall nitrification without nitrite accumulation was observed in FBR B during continuous incubation. However, in term of ammonia (AOR) and nitrite oxidation rates (NOR) that were determined in batch incubations, FBR A revealed a higher AOR and NOR of 6 and 7 mg N L^{-1} h^{-1}, compared to the AOR and NOR of 5 and 5.9 mg N L^{-1} h^{-1} in FBR B, respectively.

A better homogenization by external recirculation increased the AOR in FBR A more than twofold and the NOR more than 1.5-fold, whereas the AOR and NOR in FBR B was not improved.

For studies of the effect of fluctuating salinity on the nitrification, polyethylene/clay sinter lamellas or porous ceramic rings from FBR A and B, respectively, were used as a source of immobilized nitrifiers in smaller fixed bed reactors. Salt concentrations were decreased from 3.5% to 0.03% via 2, 1 and 0.5% and increased from 3.5% to 5, 7 and 9%.

A similar result for AOR and NOR during batch incubation was obtained for both substrata in FBR A and FBR B. The salinity changes influenced more the nitrite oxidizing bacteria (NOB) than the ammonia oxidizing bacteria (AOB) during continuous operation, which is evidenced by accumulation of nitrite in the beginnings of the phase changes.

Measurement of oxidation rates during changing conditions show that the NORs were always significantly higher than the AORs in all reactors, especially in initial phase. However, NORs were more sensitive to the salinity fluctuation than AORs, especially at lower salinity, but still stayed higher than AORs. The AORs remained constant for 0.5–3.5% NaCl and dropped to 70% and 68.5% after the salt concentration was brought to 0.034 or 5%, respectively. The NORs decreased significantly to 62% and 87.5% of initial rates after the salt concentration was changed to 2 and 5%, respectively.

To evaluate the effect of salt fluctuation on nitrification by non-halotolerant nitrifiers, reactors with fresh polyethylene/clay as supporting material were inoculated with water samples taken from a ''Brackwasser''-location at the North Sea and were continuously run. The salt content in medium was 0% at the start and was then increased up to 10.5%.

Increases of the salt concentration in a non-salt-adapted FBR suppressed more to NOR than AOR. When salinity was increased from 0.03% to 0.5, 1 and 2%, AORs remained constant. The AORs were approx. 90% of initial rates after the salt concentrations were increased stepwise to 3.5%, whereas the NORs decreased to 85, 52 and 36% of initial rate after salinity was increased to 1, 2 and 3.5%, respectively.

A temporary salt inhibition, however, was obtained in all FBRs that were either inoculated with seawater or fresh water.

The effect of temperature, ammonia and nitrite on nitrification was investigated in batch assays with biofilm-nitrifiers attached to porous ceramic rings. The inhibition concentration IC_{50} for ammonia on AOR was about 1 g N L^{-1} (representing a free ammonia concentration FA of 54 mg N L^{-1}) and on NOR 5 g N L^{-1} (representing a FA of 70 mg N L^{-1}). The IC_{50} of nitrite on AORs was about 0.027 g N L^{-1} (representing a FA of 0.0021 mg N L^{-1}), but no inhibition effect of nitrite on NOR was found.

A sudden temperature increase of 10°C from 22.5 to 32.5°C resulted in an improvement of the AOR and the NOR by a factor of 1.036 and 1.026 per °C, respectively.

Zusammenfassung

Zwei Festbettreaktoren zur halophilen Nitrifikation von Abwässern, entweder mit Polyethylen/Blähton Sinterlamellen (FBR A) oder porösen Keramikringen (FBR B) als Trägermaterialien wurden über lange Zeit zur Behandlung von synthetischem Salz- und Ammonium-haltigem Abwasser kontinuierlich betrieben. Halophile Nitrifikanten wurden aus Meerwasser vom Hafen Büsum, aus Estuarienschlamm von Stankt Peter-Ording bzw.dem Kläranlagenauslauf der Stadt Norden ins Meer angereichert. Als Inokulum für die beiden Reaktoren diente die Anreicherungskultur vom Hafen Büsum. Die Leistung der halophilen Nitrifikation wurde unter verschiedenen Reaktorbedingungen, wie z.B. unterschiedlichen Ammonium-Raumbelastungen (ALRS), unterschiedlichen Salzkonzentrationen von 0 – 10 %, unterschiedlichen pH-Werten und für verschiedene Strömungsgeschwindigkeiten des Abwassers im Reaktor bestimmt.

Bei allen Experimenten wurde auch für Extremwerte der obigen Parameter kein dauerhafter Schaden für die Nitrifikanten auf dem Trägermaterial festgestellt. Eine bessere Gesamtnitrifikation ohne Nitrit-Akkumulation wurde im FBR B mit Polyethylen/Blähton Sinterlamellen während des kontinuierlichen Betriebes beobachtet. Wenn jedoch die Ammonium- (AOR) und Nitrit-Oxidationsraten (NOR) in Batch-Inkubationen bestimmt wurden, zeigte die Anreicherungskultur aus dem FBR A auf den porösen Keramik-Ringen eine höhere AOR und NOR von 6 bzw. 7 mg N L^{-1} h^{-1} im Vergleich zur AOR und NOR von 5 und 5,9 mg N L^{-1} h^{-1} im FBR B mit PE/Blähton-Sinterlamellen.

Eine höhere Strömungsgeschwindigkeit (bessere Homogenisierung) durch externe Abwasser-Rezirkulation erhöhte die AOR im FBR A um mehr als das Doppelte und die NOR um mehr als das 1,5-fache, während die AOR und NOR in FBR B nicht erhöht war.

Zu Untersuchungen über die Auswirkungen schwankender Salzgehalte auf die Nitrifikation wurden PE/Blähton Sinterlamellen aus FBR A oder poröse Keramikringe aus FBR B in kleinere Festbettreaktoren als Quelle von immobilisierten halophilen Nitrifikanten eingesetzt. Die Salzkonzentrationen in drei parallel betriebenen Reaktoren wurden von 3,5 % (Ausgangskonzentration) auf entweder 2, 1 oder 0,5 % in je einem Reaktor und dann in allen Reaktoren auf 0,03 % abgesenkt. Sodann wurde die Salzkonzentration in den Reaktoren wieder auf 3,5 % Salzkonzentration und von da auf entweder 5, 7 und 9 % angehoben. Nur bei 7 und 9 % war die Nitrifikation total gehemmt, erholte sich aber wieder nach Rücknahme der hohen Salzkonzentration auf 3,5 %. Ein ähnliches Ergebnis für AORs und NORs während der Batch-Inkubation von bewachsenem Trägermaterial wurde für beide Substrate in FBR A und B gefunden. Die Änderung des Salzgehalts zu niedrigeren und besonders zu höheren Werten beeinflusst die Nitrit-oxidierenden Bakterien (NOB) mehr als die Ammonium-oxidierenden Bakterien (AOB) im laufenden Betrieb, was sich durch die z.T. sehr hohe Akkumulation von Nitrit zu Beginn der Veränderungsphase äußert.

Die Messungen von AORs und NORs während veränderter Rahmenbedingungen zeigten, dass in allen Reaktoren für fast alle Bedingungen die NOR immer höher als die AOR war. Die NOR war nur in der Anfangsphase nach Salz- oder pH-Wechsel kurzzeitig inhibiert und Nitrit reicherte sich an. Die Nitrit-oxidierenden Bakterien reagierten empfindlicher auf einen schwankenden Salzgehalt als die Ammonium-oxidierenden Bakterien. Die AORs blieben für Salzgehalte von 0,5-3,5 % konstant und sanken auf 70 % bzw. 68,5 % nachdem die Salzkonzentration auf 0,034 % abgesenkt oder auf 5 % erhöht wurde. Die NORs wurden deutlicher gehemmt und gingen auf 62 % und 87,5 % der ursprünglichen Rate zurück, nachdem der Salzkonzentration von 3,5 % auf 2 % abgesenkt oder auf 5 % erhöht wurde.

Um die Wirkung des schwankenden Salzgehalts auf die Nitrifikation durch halotolerante Nitrifizierer zu bewerten, wurden Reaktoren mit frischen Polyethylen/Blähton als Trägermaterial mit Wasserproben aus einer "Brackwasser"-Lage an der Nordsee geimpft und kontinuierlich betrieben. Der Salzgehalt im Medium lag zu Beginn bei 0,03 % und wurde dann bis zu 10,5 % erhöht.

Wird die Salzkonzentration in FBRs mit nicht-halophilen Inokula erhöht, wird die NOR viel stärker als die AOR unterdrückt. Bei einer Erhöhung des Salzgehaltes von 0,03 % auf 0,5, 1 und 2 % blieb AOR konstant. Die AOR betrug ca. 90 % der ursprünglichen Rate wenn der Salzgehalt schrittweise auf 3,5 % erhöht wurde. Die NOR sank dabei auf 85, 52 und 36 % der ursprünglichen Rate für Salzgehalte von 1, 2 und 3,5 % ab. War aber immer noch vergleichsweise gut, weil die Absenkung von einem höheren Niveau startete.

Eine vorübergehende Salzhemmung der NORs wurde jedoch in allen FBRs erreicht, die entweder mit Meerwasser oder Süßwasser als Inoculum beimpft wurden.

Die Wirkung der Temperatur, von Ammonium und Nitrit auf die Nitrifikation wurde im Batch-Versuch mit Biofilm-Nitrifikanten, die auf die Keramikringe aufgewachsen waren untersucht. Die Hemmkonzentration IC_{50} für die AOR beträgt ca. 1 g N L^{-1} für Ammonium und 5 g N L^{-1} für Ammonium für die NOR. Die Hemmkonzentration IC_{50} von Nitrit für die AOR betrug etwa 0,027 g N L^{-1}, jedoch wurde keine Hemmung von Nitrit, auch von hohen Konzentrationen der NOR gefunden.

Ein Temperaturanstieg um 10° C von 22,5° C auf 32,5° C führte zu einer Verbesserung des AORs und NORs um einen Faktor von 1,036 und 1,026.

Contents

List of figures

List of Tables

Abbreviations

ALR : Ammonia Loading Rate

AOB : Ammonia Oxidizing Bacteria

AOR : Ammonia Oxidation Rate

ATU : Allylthiourea

BOD : Biological Oxygen Demand

CANON : Completely Autotrophic Nitrogen removal Over Nitrite

COD : Chemical Oxygen Demand

CSTR : Continuous flow Stirred Tank Reactor

EPS : Extracellular Polymeric Substances

FA : Free Ammonia (NH_3)

FBR : Fixed-Bed Reactor

- FBR A : Polyethylene/clay sinter lamellas, basic reactor
- FBR B : Ceramic rings, basic reactor
- FBR C, D and E : Ceramic rings from FBR B for changing NaCl, with a starting concentration of 3.5% salt
- FBR F, G, H and I : Polyethylene/clay sinter lamellas from FBR A for changing NaCl, with a starting concentration of 3.5% salt
- FBR J, K, L, M and N : Fresh Polyethylene/clay sinter lamellas for changing NaCl, with a starting concentration of 0.03% salt
- FBR O, P and Q : Polyethylene/clay sinter lamellas from FBR A for effect of a sudden or gradual increase of the ammonia concentration

FNA : Free Nitrous Acid (HNO_2)

NAR : Nitrite Accumulation Rate

NOB : Nitrite Oxidizing Bacteria

NOR : Nitrite Oxidation Rate

SBR : Sequencing Batch Reactor.

SHARON : Single reactor system for High activity Ammonia Removal Over Nitrite

SRT : Sludge Retention Time

Chapter 1

INTRODUCTION

1.1 Background

Eutrophication describes a condition of water bodies having high nutrient contents especially of nitrogen and phosphorus compounds. An excessive content of nutrients stimulates biomass formation by algal blooms, which may excrete very toxic biocides and during decay can lead to oxygen starvation in water ecosystems. Oxygen deficiency leads to a reduced biodiversity of the macro- and microorganism populations. Fishes suffocate, the fish production is drastically diminished and the use of the water bodies is highly restricted.

Eutrophication becomes a serious problem in many fresh water habitats and coastal areas throughout the world (UNEP 2006). Within 10 years, from 1987 to 1997, Germany successfully decreased 25 % of the total nitrogen emission into the water body of catchment areas. Nevertheless, the international goal of reducing nutrient emissions by 50 % from 1985 to 1995 could not be reached (Behrendt et al. 2002). All states bordering Germany also missed this goal. Most parts of the German Bight of the North Sea and the Baltic Sea are classified as problem areas with a bad eutrophication status (OSPAR commissions 2008 and HELCOM 2009).

With regard to sources of pollution of natural water resources, point and non point, diffuse pollution of water bodies with nutrients could be distinguished. Failure to remove nitrogen from point sources such as wastewater treatment plants clearly contributes to eutrophication problems. Nitrogen removal from domestic and industrial wastewater after the removal of carbon compounds by nitrification and denitrification has been successfully engineered in the last decades. There are, however, sources of wastewater with specific characteristics, such as a high salt content, where classical nitrogen removal processes face several problems. Industries generating saline wastewater are for example fish or seafood industry, usually located at the coast, the leather or gelatin industry and power generation plants. Using sea water as toilet flushing water in dual wastewater systems for coastal cities, such as for instance Hong Kong generates domestic wastewater with a high salinity and requires halophilic or halotolerant microorganisms for purification.

A high content of salt in wastewater leads to a reduced osmotic pressure in and a concentration increase of the cytoplasm. Under "normal" low-salt conditions the cytoplasm membrane of wastewater bacteria is permeable for water and this causes a moderate internal cell pressure, which must be counteracted by the cell wall. If bacteria live in a salt-free surrounding, water intrudes into the cytoplasm and leads to a high internal pressure and finally to cell rupture. On the other hand, microorganisms that live in a high salinity environment must maintain the intracellular water level high enough for cell activity, otherwise osmolysis will prevent metabolic activity. Such bacteria can be found in naturally saline environment such as seawater/mud and they might catalyze the nitrogen removal process in saline wastewater.

Since the nitrifying bacteria that carry out ammonia and nitrite oxidation are autotrophic and thus slow growing, nitrification often becomes the limiting process for biological nitrogen removal. The growth rate of nitrifying bacteria is almost a factor of 10 smaller than that of heterotrophic carbon removing bacteria. For growth of nitrifiers in a saline environment, the bacteria need extra energy for carbon dioxide fixation and to maintain an intracellular osmotic pressure. As a consequence of the necessity to maintain a minimal intercellular pressure, all bacteria that live in saline environment have less energy available for growth than those which live in a sweet water environment.

Biological removal of pollutants during wastewater treatment is principally possible by suspended bacteria or by biofilm bacteria that grow attached to surfaces. For biological processes that must be carried out by slow growing bacteria, such as nitrification, attached growth on support materials enables the microorganisms to form biofilms and to reach high cell densities by carrier-supported immobilization..

Advantages of attached growth and biofilm formation for treating wastewater can be summarized as follows:

- A high population density can be maintained because bacteria attach actively onto support material and thus are not washed out with the effluent.
- An increasing system performance can be achieved due to the existence of a high density of biomass.
- High shock loading resistance and better recovery from shock loadings are a result of a protector function of presumably the extracellular polymeric substances (EPS) that keep together the biofilm.
- Returning of activated sludge to increase activity as in suspended growth reactors is not needed so biofilm reactors have reduced costs of operation.

There are also some disadvantages of attached growth processes which are:

1. Transport limitations, for instance mass transfer of oxygen or substrates through the hydrated EPS layer may limit microbial growth at the base.
2. Risk of clogging when not properly designed and operated.
3. Difficult evaluation of kinetic processes due to a complex interaction between biofilm and liquid.
4. No uniform distribution of substrates as well as of the biomass population due to a difficult mixing system.

The growth development of microorganisms that tend to attacht to a support material is influenced by several factors, inculuding pore and surface.

characteristics of the support material. The materials, which could be used as a support material are characterized as follows:

a. Inert material : Physical and biological processes in a reactor do not corrupt the material and vice versa
b. Surface roughness: The roughness represents amount and size of crevices, where microorganisms could initially growth without disturbances by shear forces
c. A reactor filled with a high porosity support material results in high void ratios in the reactor and this could reduce the clogging risk
d. Supporting media having high specific surface area provide more space for bacterial growth

1.2Objectives

The main objective of this research was to investigate the feasibility of fixed-bed reactors that contain porous ceramic rings or Pelia polyethylene/clay mats as substratum for halophilic or halotolerant nitrifying bacteria. Specific aspects of this work were:

1. To find appropriate inocula for nitrification of saline wastewater
2. To observe the long term performance of nitrification in fixed bed reactors under a changing ammonia loading rate (ALR), at changing pH and with external recirculation
3. To investigate and compare the influence of varying salt concentrations an ammonia and nitrite oxidation rates, and the performance of fixed bed reactors inoculated with a micro flora either from fresh or sea water
4. To assess the effect of substrate concentration on ammonia and nitrite oxidation rates by nitrifiers that were immobilized in a biofilm and on the process stability of fixed bed reactors treating saline wastewater.

5. To determine the ammonia and nitrite oxidation rates of nitrifiers in biofilms under different temperature.

Chapter 2

LITERATURE REVIEW

Microorganisms must be supplied with energy and carbon sources and nutrients such as nitrogen-, phosphorus- and sulfate-compounds, as well as growth factors such as zink, manganese and nickel, among many others. Nitrogen is known as an essential building block in the synthesis of protein. The forms of nitrogen in wastewater are ammonia (NH_4^+), nitrite (NO_2^-), and nitrate (NO_3^-) ions and organic nitrogen, determined as total nitrogen The organic fraction of nitrogen consists of a complex mixture of compounds including amino acids, amino sugars, and proteins (polymers of amino acids), which readily converted to ammonia through degradation of the carbon skeleton by microorganisms in the aquatic environment (Metcalf and Eddy 2003).

Nitrogen removal during wastewater treatment is necessary to avoid:

- Oxygen depletion of receiving water bodies
- Eutrophication of receiving surface water
- Effect of ammonia, nitrite and nitrate on receiving water with respect to fish toxicity
- Inefficiency of chlorine disinfection for water reuse application (Metcalf and Eddy 2003, Ahn 2006)

The nitrogen concentration of wastewater varies broadly depending on activities generating the wastewater. Nitrogen compounds can be removed from wastewater by a variety of physicochemical processes such as air or steam stripping, ion exchange, and biological processes such as nitrification and denitrification. Biological nitrogen removal has been widely applied due to its effectivness and inexpensive process operation (Ahn 2006)

2.1 Nitrogen removal process

2.1.1 Processes for biological nitrogen removal

2.1.1.1 Conventional biological nitrogen removal

Biological nitrogen removal is usually achieved by a sequence of nitrification and denitrification processes. During nitrification ammonia is biologically oxidized to nitrate via nitrite which is then reduced to nitrogen gas during the denitrification process, as shown in the Figure 2.1.

$$NH_4^+ \longrightarrow NO_2^- \longrightarrow NO_3^- \longrightarrow NO_2^- \longrightarrow N_2$$

Nitritation Nitratation Denitratation Denitritation

Nitrification Denitrification

Figure 2.1 Nitrification and denitrification processes

Nitrification is conducted in two sequential oxidative stages: ammonia oxidation to nitrite (nitritation) and nitrite oxidation to nitrate (nitratation) with oxygen. Each stage is performed by different bacterial genera that are for instance *Nitrosomonas, Nitrosococccus* for nitritation and *Nitrobacter, Nitrospira* for nitratation. The nitrifiers use ammonia or nitrite as an energy source, oxygen as an electron acceptor and carbon dioxide as a carbon source.

Equations for nitritation, nitratation and total oxidation generating energy are as follows (Metcalf and Eddy 2003).

Eq. 2.1: $2NH_4^+ + 3O_2 \rightarrow 2NO_2^- + 4H^+ + 2H_2O$

Eq. 2.2: $2NO_2^- + O_2 \rightarrow 2NO_3^-$

Eq. 2.3: $NH_4^+ + 2O_2 \rightarrow NO_3^- + 2H^+ + H_2O$

The biomass synthesis reaction in nitrification is represented as follows:

Eq. 2.4: $4CO_2 + HCO_3^- + NH_4^+ + H_2O \rightarrow C_5H_7O_2N + 5O_2$

The overall synthesis and oxidation reaction in nitrification can be represented as follows:

$$Eq.\ 2.5:\ NH_4^+ + 1.863O_2 + 0.098CO_2 \rightarrow 0.0196C_5H_7NO_2 + 0.98NO_3^- + 0.0941H_2O + 1.98H^+$$

The chemical formula $C_5H_7NO_2$ represent the synthesized bacterial cells.

According to the above equation for each g of ammonia (as N) oxidized, 4.25 g of O_2 are utilized, 0.16 g of new cells are formed, 7.07 g of alkalinity (as $CaCO_3$) are consumed and 0.08 g of inorganic carbon are required for the formation of new cells.

Denitrification is accomplished in two sequential reductions under anoxic condition by a heterotrophic bioconversion process. Nitrate is reduced to nitrite (denitratation), and then the nitrite is reduced to nitrogen gas (denitritation). During denitrification, microorganisms utilize nitrite and or nitrate as electron acceptors and organic matter as carbon and energy source. A variety of carbon sources, such as methanol, acetate, glucose and ethanol can serve for denitrification. However, because of the lowest prize, methanol (CH_3OH) is used widespread. Combined dissimilation- synthesis equations for denitrification using methanol as an electron donor are as follows (Metcalf and Eddy 2003).

$$Eq.\ 2.6:\ 5CH_3OH + 6NO_3^- \rightarrow 3N_2 + 5CO_2 + 7H_2O + 6OH^-$$

2.1.1.2 Alternative biological nitrogen removal

Nitrification and denitrification, known as conventional biological nitrogen removal processes, are proceeding slowly and are relatively expensive, referring to energy requirement for aeration, the requiring alkalinity and eventually an external carbon source.

In the past few years, several alternatively biological nitrogen removal processes have been developed, including partial nitrification, denitritation, anaerobic ammonia oxidation (the Anammox process), and its combined systems. Utilization of oxygen, alkalinity and organic compounds for alternative nitrogen removal processes were summarized by (Ahn 2006) in Table 2.1.

Table 2.1 Comparison of various nitrogen removal processes

Reaction	First phase		Second phase		Source
	Oxygen [i]	Alkalinity [ii]	Alkalinity [ii]	Organic [iii] **	
Nitrification-Denitrification	4.57	7.14	3.57*	3.7	a
Nitritation – Denitritation	3.43	7.14	3.57*	2.3	b
Partial nitritation – Anammox	1.71–2.06	3.57	0.24	-	c
CANON	1.94	3.68	-	-	d

[i] : (g O_2/g N); [ii] = (g $CaCO_3$/g N); [iii] = (g COD/g N)
* : Alkalinity production in heterotrophic denitrification
** : Based on methanol
Source : a, b : Rittmann & McCarty 2001 in Ahn 2006; c : Van Dongen et al., 2001 in Ahn 2006; d : Sliekers et al., 2003

According to table 2.1, conventional ammonia removal needs clearly more oxygen, alkalinity and organic substances, which lead to a more expensive process.

- *Nitritation and denitritation*

In this method, called also partial nitrification, ammonia is removed to nitrogen gas via nitrite. No nitrite oxidation to nitrate and no nitrate reduction to nitrite should occur, so that oxygen and organic requirements can be essentially reduced (Hellinga et al., 1998). The methods proceed practically in a single reactor or in two reactors. In a single reactor, such as in the Sharon (Single reactor system for High activity Ammonia Removal Over Nitrite) process, an intermittent aeration system is applied and nitratation is avoided by adjusting pH, the solid retention time (SRT) and dissolved oxygen (DO) to depress the activity of nitrite nitrifiers.

In the Sharon process with two reactors, aerobic conditions for the first reactor are necessary and the second reactor must be run under anaerobic conditions. Beside of that, environmental factors in the first reactor such as, pH, DO, temperature are adjusted to suppress the nitrite oxidizing bacteria. The stoichiometry of an ideal Sharon reaction is represented by Eq. 2.7 – 2.9.

Eq. 2.7: $2NH_4^+ + 3O_2 \rightarrow 2NO_2^- + 4H^+ + 2H_2O$

Eq. 2.8: $2NO_2^- + 4.8$ g COD $+ 2H^+ \rightarrow N_2 + 1.8$ g sludge

Eq. 2.9: $2NH_4^+ + 3O_2 + 4.8$ g COD $\rightarrow N_2 + 2H^+ + 1.8$ g sludge

- *Two-in-series reactors for partial nitritation—Anammox process*

Beside ammonia oxidation under aerobic conditions, it was recognized that ammonia could be also oxidized under anaerobic conditions. Using nitrite as electron acceptor, anammox bacteria can oxidize ammonia to nitrogen gas, as outlined in the following equation (Van de Graaf et al., 1995).

Eq. 2.10: $NH_4^+ + NO_2^- \rightarrow N_2 + 2\ H_2O$

Based on equation 2.10, 1 mol ammonia needs 1 mol nitrite or a ratio of 1:1of ammonia:nitrite should be maintained to support the anammox process. Under practical conditions nitrogen compounds can be eliminated by anaerobic ammonia oxidation. However, ammonia must partly be pre-oxidized to nitrite (about 50% of ammonia), but not to nitrate, before feeding into the anammox process. So in this two-in-series partial nitritation-anammox process two reactors are required, where in the first reactor partial nitrification (conversion of about 50% of ammonia to nitrite) proceeds and in the second reactor the anammox process takes place.

- *CANON (Completely Autotrophic Nitrogen removal Over Nitrite)*

A further development of the two-in-series partial nitritation—anammox process is the Canon process. The concept of the Canon process is to combine partial nitritation and anammox in one fixed-bed reactor (Van Loosdrecht et al., 2004). In a biofilm (attached growth conditions) the nitrifiers oxidize partly ammonia to nitrite at the surface, consume the oxygen and then create anoxic conditions farther inside the biofilm for ammonia and nitrite disproportionation which is needed for the anammox process. The operational conditions of the Canon process are relatively sensitive, concerning especially the biofilm thickness, the dissolved oxygen concentration, the nitrogen-surface load and the temperature.

2.1.2 Nitrification biokinetics

Biological processes in wastewater treatment depend on substrate utilization and microbial growth. The basic principal conditions for biological reactions and growth of microorganisms must be obeyed for designing and operating a denitrification treatment system (Metcalf and Eddy 2003). The rate of oxidation (i.e utilization) of substrates and production of biomass is regulated by the kinetics of microbial growth under the operation conditions.

The kinetics of the nitrification process has been the subject of several studies (e.g. Dahl et al. 1997, Dincer and Kargi 2000) and the Monod Equation (Equation 2.11) is usually used to describe the growth rate of the bacteria as a function of substrate concentration. (Grady et al. 1999)

Eq. 2.11: $$\mu = \frac{\mu_{max} S}{K_S + S}$$

μ = growth rate (g new cells. g cells^{-1}.d^{-1})
μ_{max}= maximum growth rate (g new cells. g cells^{-1}.d^{-1})
S = Concentration of limiting substrate (g L^{-1})
K_s = Monod Constant, that is the substrate concentration when the growth rate is equal to the half maximum growth rate (g L^{-1})

The equation is the same form used to estimate Michaelis-Menten enzyme kinetics, but it was transferred by Monod to describe the specific growth rate of bacteria, when the limiting substrate was available for the microorganisms in a dissolved form and the time interval was very short.

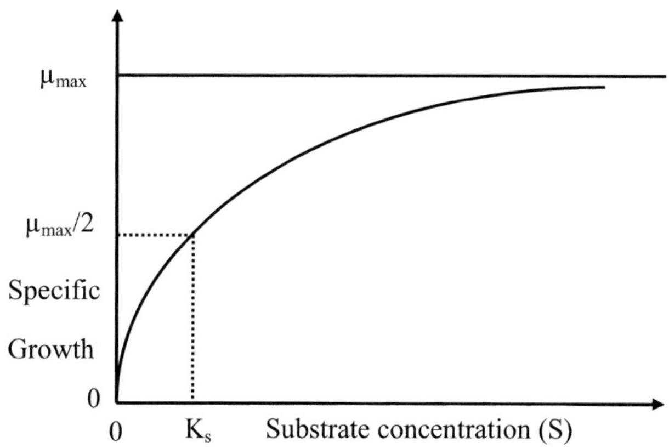

Figure 2.2 Specific growth rate of as a function of substrate concentration as depicted by the Monod equation

Data to obtain the Monod Equation are shown in Figure 2.2; the maximum substrate utilization rate occurs at high substrate concentration and when the substrate is being used at its maximum rate, the bacteria are also growing at their maximum rate. Thus, the maximum specific growth rate of the bacteria also takes place at high substrate concentration.

2.1.3 Factors affecting nitrification

There are many chemical and biological factors that can affect the growth and then influence the performance of nitrifying bacteria. The most significant factors can be classified into three major categories, as summarized by Chen et al. (2006).

1. The first category includes those that affect the biochemical process of the microbes such as pH, temperature and salinity.
2. The second category includes those that affect the supply of nutrients to the biofilm such as substrate concentration, dissolved oxygen (DO) and mixing regime.

3. The third category includes those that have impact upon both growth and nutrient supply, such as the competition for either essential nutrients or space, represented by the level of organics.

2.1.3.1 *Substrate concentrations*

The concentration of ammonia as the substrate of nitrification is the important factor for growth. Two essential questions are related to nitrification processes: the minimum ammonia concentration for nitrification to proceed properly and the maximum concentration at which ammonia and product concentrations are still lower than toxic concentration level.

The substrate should refer to the Monod Equation to achieve a high substrate utilizing without causing toxic effects for microorganisms.

Eq. 2.12:

$$FA\left(NH_3\ mg\ L^{-1}\right)=\frac{17}{14}x\frac{[NH_4^+]\ x\ 10^{pH}}{e^{\left[\frac{6344}{(273+t)}\right]}+10^{pH}}$$

Eq. 2.13 :

$$FNA\left(HNO_2\ mg\ L^{-1}\right)=\frac{46}{14}x\frac{\left[NO_2^-\right]}{e^{\left[\frac{-2300}{(273+t)}\right]}\ x\ 10^{pH}}$$

The inhibition effect of ammonia and nitrite was attributed to the concentrations of free ammonia (FA) or free nitrous acid (FNA) by Anthonisen et al. (1976), which could be calculated with Equation (2.12) and (2.13).

Since only FA and FNA reveal a high permeability through cell membranes several hypotheses have been proposed to interpret the effects of FA and FNA on microbial cultures, such as a decreasing intracellular pH, and thus interference with the trans-membrane pH gradient required for ATP synthesis (Anthonisen et al., 1976).

2.1.3.2 *Dissolved oxygen*

Oxygen is a requirement in ammonia oxidation. According to equation 2.1 and 2.2, the theoretical oxygen requirements are 3.43 mg for oxidation of 1 mg

NH_4^+–N to nitrite and 1.14 mg for oxidation of 1 mg NO_2-N to nitrate.Effects of the DO concentration on nitrification rates have been observed by several authors in both, attached and suspended growth systems. (Sharma and Ahlert 1977, Stenstrom and Poduska 1980, Beccari et al. 1992).

Most of the investigations concluded that a DO concentration above 5 mg L^{-1} should be maintained for nitrification. It is found that nitritation takes place at an optimum DO concentration above 1 mg L^{-1}, whereas nitratation works optimal at DO concentrations above 2 mg L^{-1} (Stenstrom and Poduska 1980). Most of the above results are, however, based on research in suspended growth process.

Because of the nature of diffusion limitation in fixed films, however, oxygen limitation in biofilms for nitrification can be significantly different compared to oxygen availablility in suspended growth processes (Zhu and Chen 2002). It is proposed that a higher DO concentration in fixed film processes should be maintained than in suspended growth processes. A microelectrode and a micro slicing technique enabling the measurement DO concentration profiles within biofilms would be required means for experimentation.

2.1.3.3 Turbulence

A limiting factor for nitrification rates in biofilms or flocs would be mass transfer of substrates and nutrients. The fluxes of nutrient masses, which are affected by the thickness of water films, determine the efficiency of biological processes in biofilms. As turbulence in water affects the thickness of the water films, the water flow behavior has an effect on the flux of substrates from the bulk liquid into the biofilm as well as on the nitrification rate.

De Beer et al. (1996) reported that an increasing flow velocity, which led to higher turbulence, decreased the thickness of the "mass transfer boundary laye" on the biofilm. Zhu and Chen (2001) investigated the relationship between total ammonia nitrogen removal rates and the Reynolds number (R_e) in a steady-state nitrification fixed-bed reactor and reported that substrate consumption of the nitrifying biofilm during steady state followed a two-step process. The first was an external mass transfer or transport of substrates in the medium and the second was an internal mass diffusion or transport of substrates within the biofilm.

2.1.3.4 Organics

The organic compounds in wastewater exists in dissolved and particulate form and could influence the nitrification process. The most important impact of organics on nitrification is due to the fact that these compounds serve also as substrates for fast-growing heterotrophic bacteria. Thus, heterotrophic bacteria consume oxygen for oxidizing the organic compounds and, consequently, an additional oxygen demand is necessary.

Besides that, the heterotrophic bacteria compete with nitrifying bacteria for optimal growth space on the carrier materials of fixed.bed reactors (Ohashi et al. 1995, van Benthum et al. 1997).

2.1.3.5 Temperature

Temperature effects on biological reaction rates can be quantified using the van't Hoff-Arrhenius Equation. A higher temperature enhances biochemical bacterial processes as well as nitrification rates. Based on the equation 2.14, the reaction rate almost doubles by increasing the temperature by 10°C. Nitrification kinetics seems to be very sensitive to elevated temperatures.

Eq 2.14:

$$k = A.e^{\left(-\frac{E_a}{RT}\right)}$$

k = rate coefficient
A = a constant
e = the base of natural log
E_a = the activation energy (kJ mol^{-1})
R = the universal gas constant (8.314 x 10^{-3} kJ $mol^{-1}K^{-1}$)
T = the temperature (in kelvin).

With increasing temperature nitrification rates increase to an apparent maximum at 30-37.5°C whereas beyond 37.5°C the overall rates decrease (Fontenot et al. 2007).

In biofilm systems, the effect of temperature on nitrification kinetics are more complicated to describe than in suspended cultures due to other factors such as limitations of oxygen and reduced mass diffusion (Fdz-Polanco et al. 1994, Chen et al. 2006).

Ammonia oxidizing bacteria (AOB) and nitrite oxidizing bacteria (NOB) responded differently to temperature changes: an increase of the incubation temperature increased the ammonia oxidation rate (AOR) more than the nitrite oxidation rate (NOR). After an increase of the incubation temperature (Kim et al. 2008) nitrite build-up and residual ammonia were detected in the effluent of continuous reactors.

2.1.3.6 pH

Each enzyme in the metabolism of microorganisms works optimally at a certain pH, with decreasing activities at pH values above and below that point.

Many studies were conducted to investigate the effect of the pH on nitrification. The results vary widely, when different nitrification methods, reactors and inocula (bacteria) were compared (Biesterfeld et al. 2001). The optimum pH for nitrification ranges from 7.0 to 9.0 with the optimum pH for *Nitrosomonas* sp. ranging from 7.2 to 8.8 and for *Nitrobacter* sp. from 7.2 to 9.0.

Mechanisms of the effect of pH on nitrification were summarized by Villaverde et al. (1997):

- activation– deactivation effects of nitrifying bacteria, corresponding to the fact that enzymes work optimally only at a certain pH.
- nutritional effects connected with alkalinity, corresponding to the fact that autotrophic bacteria utilize inorganic carbon such as $NaHCO_3$.

- inhibition effect through free ammonia and free nitrous acid, corresponding to the fact that the concentration of free ammonia and free nitrous acid depend on pH.

2.1.3.7 Alkalinity

The alkalinity is necessary during conversion of ammonia to nitrate as shown in Equation 2.4. Beside its function as a nutrient element for nitrifying bacteria, alkalinity in the form of carbonate and bicarbonate provides the buffering capacity to prevent pH changes in the nitrification process. So, the impact of alkalinity on the nitrification rate is related very much to the pH.

As shown in Equation 2.4, 7.01 mg alkalinity as $CaCO_3$ is needed to oxidize 1 mg NH_4^+–N.

2.1.3.8 Salinity

As biological membranes are permeable for water, salt concentrations influence osmotic pressure in intracellular compartments and lead changing activities of bacteria. Detailed effects of salinity on nitrification are discussed in section 2.3.

2.2Suspended and attached growth of bacteria

The principal biological processes used for wastewater treatment can be divided into two main categories: suspended growth and attached growth. In suspended growth processes, the microorganisms responsible for treatment of wastewater are maintained in liquid suspension by appropriate mixing methods exemplified by the activated sludge process for BOD removal and nitrification and process variations/modifications for nitrogen and phosphorus removal.

In attached growth processes, the microorganisms responsible for the conversion of organic material or nutrients are attached to an inert packing material. The organic material and nutrients are removed from wastewater flowing past the attached growth microorganism also known as a biofilm.

Common names from processes that use aerobic or anaerobic suspended bacteria (e.g. “sludge flocs”) are activated sludge process, aerated lagoons, anaerobic contact processes, CSTR, etc., and for attached growth processes trickling filters, rotating biological contactors, up flow anaerobic sludge blanket, etc.

2.2.1 Suspended growth

In reactor applying suspended growth processes such as activated sludge treatment, microorganism are maintained in suspension form with aeration providing a proper mixing. The microbial suspension generally is referred to as mixed liquor (volatile) suspended solids (Metcalf and Eddy 2003).

An important feature of the activated sludge process is the formation of floc particles ranging in size from 50 – 200 μm, which can be removed by gravity settling in a sedimentation tank.

The activated sludge flocs are often described as having two major fractions, a loosely and a strongly bound fraction, both mainly consisting of bacterial cells and extracellular polymeric substances (EPS) (Keiding and Nielsen 1997, Liao et al. 2002, Sheng et al. 2006).

In activated sludge processes where organic compounds (COD) and nitrogen are removed, nitrifiers are known to grow in dense micro colonies inside (Wagner et al. 1995, Mobarry et al. 1996, Daims et al. 2001), which seem to form the strongest fraction of the flocs (Jorand et al. 1995, Biggs and Lant 2000).

2.2.1.1 Suspended bacteria

Floc formation in the treatment of wastewater with activated sludge seems to play a very important role and deflocculated activated sludge lead to a poor activity for treatment of wastewater (Tenny and Stumm 1965 in Morgan et al. 1990). Although floc-forming bacteria and their floc-forming mechanism in activated sludge have been investigated for many years, there have been some discrepancies among the results obtained by many investigators. However, biopolymers are believed to play a central role in the floc formation. Scientists roposed that polymers excreted and disposed at the microbial surface may act as absorbents and bridges between cell surfaces and therefore initiate floc formation. The sorptive cohesion of tissue cells and colloidal reaction in biofloc systems have long been thought to be influenced by naturally occurring polymers.

The bioflocculation is essential to enable the efficient and economic operation of activated sludge wastewater treatment processes.

2.2.1.2 Transport mechanisms in suspended flocs

Activated sludge flocs can contain only aerobic or both aerobic and anoxic zones, as illustrated in Figure 2.3. Thickness of the flocs and concentration of oxygen in bulk liquid influence either one zone or the two zones.

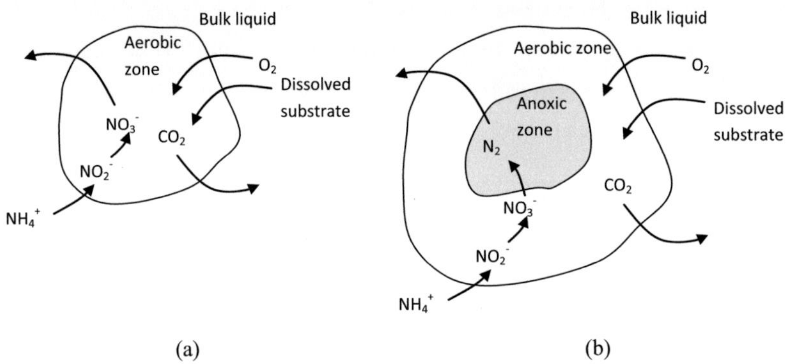

Figure 2.3 Model of activated sludge flocs with (a) only aerobic (b) aerobic and anoxic zones

In the one-zone-floc, dissolved oxygen, ammonia and dissolved substrate in the bulk liquid diffuse into the floc, and then autotrophic and heterotrophic bacteria utilize the compounds for nitrification and respiration. During nitrification, nitrite produced by ammonia nitrifiers is further oxidized into nitrate, which then is excreted and diffuses into the bulk liquid.

In the two-zone-floc, due the floc thickness and oxygen utilization by autotroph and heterotroph bacteria, oxygen cannot penetrate into inner layers of the floc, leading to anoxic zones. Nitrate produced by nitrificaton in the aerobic zone diffuses into the anoxic zone and the denitrification can start denitrification if a carbon source is available.

2.2.1.3 Nitrification in suspended bacteria

Nitrification can be accomplished in both suspended growth and attached growth biological processes. For suspended growth processes, nitrification is generally achieved along with BOD removal in the same single-sludge process, consisting of an aeration reactor, clarifier and sludge recycle system. In the reactor nitrification becomes a key design parameter. The reason for that is because the ammonia removal process is quite vulnerable due to the slow growth rate of the nitrifiers and their high sensitivity against toxic compounds (Wagner and Loy 2002).

Another system, the so-called two sludge process, consists of two aeration reactors and two clarifiers in series. Because the bacteria responsible for nitrification grow much slower than heterotrophic bacteria, systems designed for nitrification generally have much longer hydraulic and solid retention times than those designed only for BOD removal. In general, due to the vulnerability of nitrification, BOD and toxic substances are removed in the first reactor.

2.2.2 Attached growth

As explained in 2.2, microorganisms growing and attaching to an inert packing material are known as biofilm. The biological processes occur during organic material and nutrients flows past the biofilm. The biofilms exist in and on an inert packing material, which should be resistant to biological or physical corrosion processes, cheap, light and have a large surface area.

In a continuously operated fixed-bed reactor with an attached biofilm, microorganism are not washed out with the effluent, if attachment on the substratum is strong enough to withstand shear forces by the wastewater flow. So, the microorganism can be concentrated in the reactor and improve the performance of the reactor. Besides of that, a thick biofilm allows a much better resistance and recovery of microorganisms from shock load or toxic influences. The other advantages of attached growth processes at full scale are less required energy, simple operation modes and no problems with bulking sludge.

2.2.2.1 Biofilm formation and growth

In biofilm development at least four distinct stages can be distinguished, as summarized by Stoodley et al. (2002):

- reversible attachment

The individual adherent cells that initiate biofilm formation on a surface are surrounded by only small amounts of exopolymeric material and many are capable of independent movement. These adherent cells are not yet committed to the differentiation process leading to biofilm formation, and many cells may actually leave the surface. During this stage of reversible adhesion the bacteria exhibit several species-specific behaviors, which include rolling, creeping, aggregate formation, and windrow formation, before they begin to exude exopolysaccharides and adhere irreversibly.

- irreversible attachment

Once reversible attachment to a surface has been proceeded to irreversible attachment, the bacteria must maintain contact with the substratum. This change from reversible to irreversible attachment was characterized as the transition from a weak interaction of the cell with the substratum to a permanent bonding. The bacteria begin to exude exopolysaccharides to manifest the transition.

After that interactions of bacteria with one another at a surface form groups of cells, and help to strengthen the degree of attachment to the surface.

Single cells produce a polysaccharide that bonds the cells together and facilitates the formation of micro colonies and this leads to a maturation of the biofilm.

- maturation

Maturation is the next phase of biofilm development resulting in the generation of channels, pores and a redistribution of bacteria away from the substratum. In this stage, many proteins are detectable in mature biofilm samples, which represent many diverse bacteria. Varied activity is also identified in that stage, such as changing metabolism, membrane transport, secretion, adaption and protective activity.

- detachment

Detachment is a generalized term used to describe the release of cells, either individually or in groups from a biofilm or substratum. A detailed theory of detachment is explained in section 2.2.2.3.

Detached cells are believed to return to the planktonic mode of growth, thus closing the biofilm developmental life cycle. A very short schematic overview, taken from Stoodley et al. (2002), is shown in Figure 2.4.

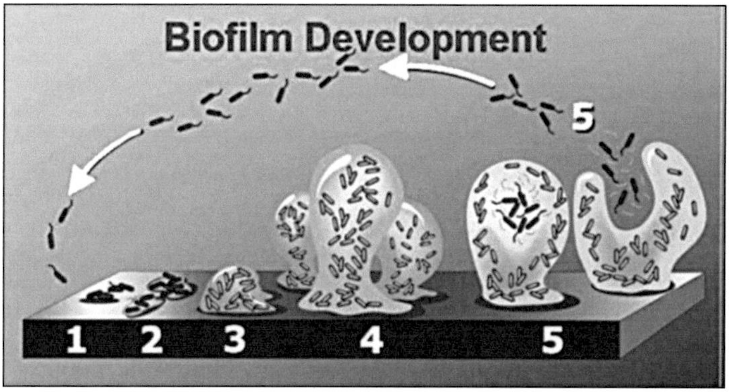

Figure 2.4 The development of a biofilm : 1. reversible attachment, 2. irreversible attachment, 3. maturation, 4. detachment, 5. return to planctonic life and start the cycle again

2.2.2.2 *Mechanisms of transport in biofilms.*

Process kinetics for suspended bacteria in an activated sludge process are based on concentrations of substrate and nutrients in the bulk liquid. By comparison, in biofilm processes, utilization of substrates and nutrients occur within outer and inner layers of attached cells.

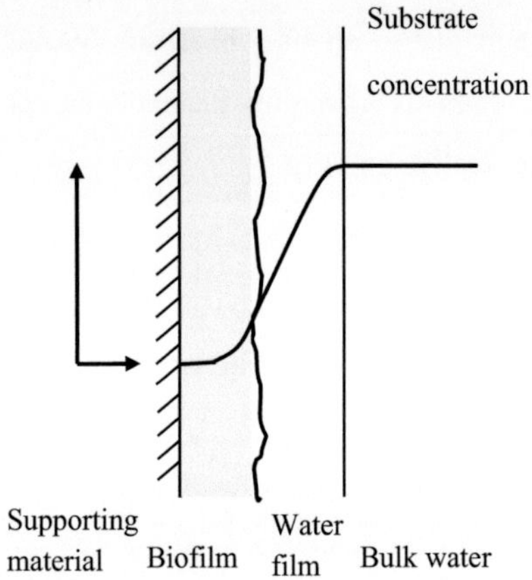

Figure 2.5 Substrate profile from bulk water through water film, biofilm towards the medium (substratum)

Nutrient and substrate concentrations vary in different lyers of a biofilm as a function of substrate transfer rates. The substrate transfer rates in a biofilm at steady state conditions can be described as a two – step process (Figure 2.5): external transfer and internal mass diffusion Zhu and Chen (2002).

- External mass transfer

It is defined as substrate or nutrient transfer from bulk water to the water-biofilm interface. A combination of molecular diffusion and convection enables the transfer (Hamdi 1995). Figure 2.5 show that the substrate and nutrient concentrations between bulk liquid and water-biofilm interface are almost equal.

- Internal mass diffusion

It is defined as substrate or nutrient transfer from the water-biofilm interface to the respective microorganisms in the biofilm. Molecular diffusion is the only mechanisms of mass transfer in this stage and is interfered by competition of different bacteria species for the same substrate. Based on figure 2.5, substrate or nutrient concentrations in biofilm, in which the microorganisms live, and in the bulk liquid is obviously different.

2.2.2.3 Factors affecting biofilm formation

The structure of a biofilm is largely heterogenic as a result of an interaction of the microorganisms with the medium and effects the biological-physical-chemical processes inside. All factors above should be considered during biofilm formation. Stoodley et al. (2002) stated that there are at least four major influences on biofilm structure:

- Geometrical characteristics of the surface of the substratum
- Characteristics of microorganisms constituting the biofilm
- Hydrodynamic conditions around the biofilm
- Nutrient availability in the liquid phase and within the biofilm
- Substratum characteristics (hydrophilic, hydrophobic reaction)

In early stages of biofilm formation substratum characteristics play an important role. Roughness of substratum promotes bacterial colonisation. A similar result is obtained by observing biofilm development during the start-up period of an expanded-bed reactor. It is hypothesized that crevices in rough surfaces can protect biofilm growth during the start period from shear forces. This enables biofilm development into next stages.

- *Biological characteristics*

Microorganisms grow under many different conditions, thus biofilm accumulation is also affected by the types of microorganisms. Microorganisms, for example, have different growth rates and biomass yields (depending in the metabolic pathways) and these will lead to a difference in biofilm structure.

Tijhuis et al. (1994) reported that nitrifying bacteria form a much denser biofilm than heterotrophic bacteria under similar reactor condition. A similar result was obtained for methanogenic and acidifying bacteria in a biofilm. The different ability of bacteria to produce extracellular polymeric substances is another important aspect of biofilm structures.

- *Hydrodynamics*

The characteristics of liquid flow pattern influence biofilm development in several ways. The mass transfer of nutrients into biofilms and of products out of biofilms is determined by the liquid flow pattern. A higher flow velocity leads to a higher mass transfer and, consequently, a faster formation of biofilms.

High shear forces at the biofilm surface could be caused by turbulent fluid flow, leading to high biofilm detachment rates.

- *Nutrient availability*

The nutrient availability is determined by the flux of substrate by diffusion towards the biofilm bacteria from the liquid environment. Nutrient availability should be analyzed together with hydrodynamics. Both factors are important for establishment of a biofilm.

At steady state and at a constant substrate loading rate, increasing fluid shear stress increases the biofilm density and decreases the biofilm thickness. An explanation of this observation is that coefficients of substrate transfer from bulk liquid to the biofilm surface increase during increasing fluid shear.

The biofilm density and its thickness increase with an increasing substrate loading rate for non-changing shear forces.

2.2.2.4 Biofilm detachment

Detachment is a late phase of biofilm development and is a generalized term used to describe the release of cells either individually or in groups (flocs) from a biofilm or substratum (Telgmann et al. 2004). It is hypothesized that biofilm "breaks" at points where the mechanical stress exceeds the biofilm mechanical strength. The mechanical stress in or at the surface of biofilms builds up due to forces acting at first on the surface as a result of the liquid flow (Stoodley et al. 1997). The mechanical strength of the biofilm is influenced by the composition and the structure of the polymeric matrix that form the biofilm. The kind of polymeric matrix is dependent on the developmental stages of a biofilm. The biofilm strength seems to correlate directly with biofilm density. High densities of biofilms can be explained by increasing fluid shear stress and substrate loading rates (Wasche et al. 2002).

Bryers (1988, cited by Telgmann et al. 2004) reported that there are different mechanisms of detachment: erosion, sloughing and abrasion. Erosions and abrasions are characterized by a continuous removal of cells or particles from the surface of the biofilm by different actions. Erosion is caused by shear forces of the moving fluid in contact with the biofilm surface, while abrasion is caused by the collision of biofilm support particles of fixed bed reactors, for instance during the regular backwashing process. Both mechanisms remove bacterial surface layers effectively from the entire surface of the substratum, whereas bacteria at the base of the biofilm are resistant and are not removed.
Sloughing is detachment of parts of a too thick by gravity forces. The more biofilm is detached the thicker the biofilm outside the substratum is. Sloughing, as erosion, is initiated by moving fluid past the biofilm and inside the cavices of the biofilm structure. During sloughing a fraction of the biofilm is removed possibly down to the substratum but detachment is not effective for the entire surface of the substratum or biofilm.

As detachment is known as an essential phase during biofilm development. Under steady state conditions the overall amount of biofilm growth equals the total amount of biofilm detachment. In other words, the total growth rate seems to be zero during steady state conditions.

2.2.2.5 *Nitrification in biofilms*

Reactors with fixed biomass have been used in wastewater treatment for the removal of ammonia. Among various types of biofilm reactors, the following are considered to be the most feasible and economical for practical applications at the present time. These are trickling filters, rotating biological contactors, biofilters and fluidized bed reactors.

In wastewater treatment, especially for domestic wastewater treatment, for the removal of organic matter and ammonia, nitrification is always preceded by organic material removal processes. The reason for that is that in the presence of high concentrations of organic compounds in a nitrification reactor growth of heterotrophic bacteria leads to a competition for oxygen and heterotrophic bacteria outcompete the nitrifying bacteria because of much shorter generation times. Autotrophic bacteria in general have low growth rates and thus heterotrophic bacteria must be grown under C-limitation to maintain a balanced parallel development.

The processes for the removal of organic matter and ammonia can be performed in following reactor systems (Boller et al. 1994).

- A single biofilm reactor: organic matter and ammonia removal take place in one reactor.
- Two separate biofilm reactors with a solids separation: the biofilm reactor for organic matter removal is followed by a biofilm reactor for nitrification and then by a sedimentation reactor.
- Two-stage treatment with two solids separation ponds: the reactor for organic matter removal is followed by sedimentation reactor. The liquid is pumped into a nitrification reactor and then into a solid separation pond to remove the nitrifying biomass.

Parameters affecting the performance of nitrifying biofilms could be divided into two categories. The first category parameters are responsible for transport and reaction processes on a microscopic level within the biofilm. The second category of parameters are related to the macroscopic level, and describe hydraulic conditions, gas exchange processes, reactor configuration and temperature conditions.

The most important parameters were summarized by Boller et al. (1994) and are listed in Table 2.2.

Table 2.2 Parameters affecting the performance of nitrifying biofilms on a biofilm oriented (microscopic) and a reactor specific (macroscopic) level

Biofilm specific parameters	Reactor specific elements
Concentration of dissolved nutrients at and in the biofilm: COD, NH_4^+, NO_2^-, NO_3^-, O_2	Kinds of reactor : completely stirred, plug flow, mixed
Concentrations of HCO_3^- and pH	Reactor hydraulics: laminar, turbulent flow
Diffusion coefficients (f (°C) for: COD, NH_4^+, NO_2^-, NO_3^-, HCO_3^-, O_2	Oxygen transfer = f (°C)
Maximum growth rates of microbial species= f (°C): Nitrosomonas, Nitrobacter	Detachment of biofilm
Saturation coefficients = f (°C) for: COD, NH_4^+, NO_2^-, NO_3^-, HCO_3^-, O_2	
Biomass densitiy and biofilm thickness	

A change of reactor operation could affect, for example, substrate concentrations in the bulk phase (macroscopic level) and then in the biofilm phase (microscopic level). Finally, a change of the biofilm phase leads to a change of biofilm activity, which is usually expressed as reaction rates. By considering the relationship between reaction rates and changes in macroscopic levels, a biofilm process can be established.

2.3 Nitrification in saline wastewater

2.3.1 Saline wastewater

Saline wastewater, which is generated by activities such as fish processing, petroleum, flue gas desulphurization (FGD) and leather industries as well as during application of seawater for toilet flushing, is characterized by the high salinity and nutrient content at the same time. Salinity and nutrient concentrations of different wastewater sources are summarized in the table 2.3. It is apparent that the concentrations vary depending on the activities as well as processes conducted in the respective industries.

The volume and concentration of wastewater from fish processing depend mainly on the raw fish composition, additives used, processing water sources and the unit processes. The main components of fish processing wastewater are lipids and protein (Palenzuela-Rollon et al. 2002) leading to the high nitrogen concentration in saline wastewater.

Table 2.3 Characterization of saline wastewater

Activities	Salt concentration (%)	Ammonia	Source
Fishery	14.6–17.9 Cl^- mg L^{-1}	0.039–1940	Aspe et al. (1997)
Tannery	2.7 (%)	1200	Lefebvre et al. (2005)
FGD	5 (%)	80	Dahl et al. (1997)
Domestic saline wastewater	0.55±0.05 (%)	130	Tang et al. (2007)
Domestic Wastewater	0.01 (%)	40	Metcalf & Eddy (2003)

Salinity has a significant chemical and physical effect on the properties of water or wastewater such as solubility of oxygen, pH as well as alkalinity.

For example, solubility of oxygen at 25°C and a air pressure of 1 bar is 8.3 mg L^{-1} in fresh water and 6.6 mg L^{-1} in saline water 3.5%. It is also known that salt tends to depress the dissociation of bicarbonate, so the carbonate system in the saline water changes. Alkalinity decreases during increases of salinity and vice versa.

2.3.2 Bacteria in saline wastewater

Microorganisms can be found over the whole range of salt concentrations from freshwater and marine to hypersaline environments with NaCl concentrations up to saturation. Thus, types of microorganisms regarded to as extreme, moderate or slight halophiles and halotolerant microorganisms could be classified based on the tolerable saline spectrum for survival. Imhoff (1986) classified the bacteria according to the salt concentration required for optimum growth as non-halophilic (grow below 0.2 M NaCl), slightly halophilic (grow at 0.2 to about 1.0–1.2 M NaCl), moderately halophilic (grow at about 1.0–1.2 to 2.0–2.5 M NaCl) and extremely halophilic bacteria (grow at 2.0–2.5 M NaCl or more).

Furthermore, "halophilic" is characterized by the obligate requirement of a high salt concentration for optimum growth, while halotolerance is described by the ability to grow at a salt concentration higher than optimum (Imhoff and Thiemann 1991). Halophilic is also characterized by the salt requirement for activity, stability and solubility of respective enzymes.

To maintain activity of halophiles ion concentrations in the cytoplasm of halophilic bacteria should be equal to the ion concentrations in surrounding liquid. Biological membranes, such as the cytoplasmic membrane are permeable for water, so that the difference ion or solute concentration causes different osmosis pressures between inside and outside membranes.

Hypertonic conditions, during which the osmotic pressure outside the membrane is higher than that inside, result in a rapid loss of water from the cytoplasm to the environment. Contrary, water flows from the environment through the cytoplasm into the cells under hypotonic conditions.

Therefore halophilic and halotolerant bacteria need to maintain their intracellular pressure at a slightly higher or the same pressure as that of their environment.

The bacteria living in saline environment exhibit modified physiological and structural characteristic. The modifications are presumably necessary to maintain their cytoplasmic pressure. Halophilic bacteria are usually elongated and swollen or show shrinkages in hypo-/hypertonic media, respectively, which shows their ability to change the cell and cytoplasmic volume.

The structural adaption concerns a change of the composition of the cell envelope and the cell membranes. To regulate the osmotic pressure in the presence of high salinity halophilic bacteria apply two kinds of strategies for osmoregulation (Galinski and Trüper 1994). The first strategy is called to "salt-in strategy". The bacteria maintain an ion concentration in the cytoplasm that is similar to that of the surrounding NaCl-containing medium, however the ions of the cytoplasm are KCl. The ion K^+ probably enters the cells passively via a uniport system and is accumulated in accordance with the size of the membrane potential (Oren 1999). In bacteria using this strategy, all enzymes and other bacterial components must be adapted to the presence of high salt concentrations and all enzymes must properly function at these high salt concentrations. The adaptations proceed over a long time. Consequently the bacteria using this strategy generally cannot survive in low salinity media.

The second strategy, the "compatible-solute strategy" is characterized by the ability of the bacteria to actively excrete NaCl and produce organic compatible solutes. The compatible solutes are described as organic osmolytes which are responsible for osmotic balance and allow enzymes to function efficiently (Galinski and Trüper 1994). Such solutes are detected in halophilic and halotolerant bacteria and comprise glycerol, arabitol, sucrose, trehalose, glycine, betain, etc.

Due to their ability to exclude NaCl from uptake and production of solutes instead, the bacteria applying the strategy do not need specially adapted proteins. Such bacteria can often live over a broad salt concentration range.

To maintain the osmotic balance between intra- and extracellular concentrations either with the salt-in or the compatible-solute strategy, bacteria need more energy. Therefore life at high salt concentration is more costly from a bio energetic point of view.

2.3.3 Nitrification in saline wastewater

Nitrogen removal of wastewater, including saline wastewater is essential to meet wastewater discharge criteria before treated wastewater is guided into a water body.

Conventional nitrogen removal processes for protein or ammonia containing saline wastewater are conducted by nitrification, followed by anoxic denitrification with addition of an external carbon source, e.g., in sequencing batch reactors (Fontenot et al. 2007).

Chemolithoautotrophic nitrification proceeds in two steps, catalyzed by phylogenetically different bacteria. Ammonia is oxidized to nitrite (nitritation) by ammonia-oxidizing bacteria, and the nitrite is further oxidized to nitrate (nitration) by nitrite-oxidizing bacteria.

Alternatively, new approaches like the completely autotrophic nitrogen removal over nitrite with nitritation of ammonia and anaerobic oxidation of ammonia with nitrite to gaseous nitrogen could be applied for treatment of saline wastewater in rotating biological contactor reactors (Liu et al. 2008).

Halotolerant or halophilic bacteria must be present to cope with the salt content of a certain wastewater. The utilization of halophilic microbial consortia or even of enrichments from non-saline ecosystems like manure, that were adapted to saline conditions, reduces the effect of salt stress on bacterial metabolism (Dincer and Kargi 2001, Antileo et al. 2002, Mariangel et al. 2008).

Salt-adapted microorganisms were also used in an anaerobic/ anoxic/aerobic system for an improved organic matter and nitrogen removal (Panswad and Anan 1999). Nitrogen removal in real or artificial wastewater in the presence of 0–6% NaCl in lab- or full-scale sequencing batch reactor (SBR) systems has been investigated (Campos et al. 2002, Fontenot et al. 2007, Huilinir et al. 2008).

With increasing salt concentrations up to 6% removal efficiencies decreased drastically in lab scale SBRs inoculated with salt-adapted, but non-halophilic, microorganisms (Intrasungkha et al. 1999, Uygur and Kargi 2004), whereas the highest nitrifying activity of a halophilic bacterial population was obtained for an in situ NaCl concentration of 28 g L^{-1} (Fontenot et al. 2007).

A negative effect of the salt concentration during nitrification was found in CSTRs treating wastewater from flue gas desulphurization or artificial wastewater (Dahl et al. 1997) or in an activated sludge system inoculated with pure cultures of *Nitrosomonas* and *Nitrobacter* species (Dincer and Kargi 2001).

In attached bacterial growth systems, nitrogen removal in the presence of high salinity has also been investigated (Vredenbregt et al. 1997, Rosa et al.1998, Gharsallah et al. 2002, Windey et al. 2005). Attached growth on surfaces of support materials has many advantages as compared to suspended growth in flocs or granules, for instance a long sludge retention time, prevention of washout of biomass and better process tability in terms of withstanding shock loadings or short-term inhibitory effects (Fitch et al. 1998, Nogueira et al. 1998). However, the application of biofilm systems for nitrogen removal could not avoid of nitrification by high salt concentration.

Fluctuations of salinity in wastewater depend on treated raw materials and different wastewater streams that may change during a day in industrial production. Thus, carbon removal and nitrification in biological wastewater treatment processes should function over a wide range of salt concentrations.

Ammonia-oxidizing and nitrite-oxidizing bacteria might respond differently to changes of environmental conditions like varying salinity. Specific inhibitors such as allythiourea and sodium azide should be applied to facilitate the separate investigation of AOB and NOB activities (Sanchez et al. 2005). A higher sensitivity of AOB to high salt concentrations of pure cultures was reported (Moussa et al. 2006). Controverse results were found by Kawasaki et al. (1996) or Vredenbregt et al. (1997).

Chapter 3

MATERIALS AND METHODS

3.1 Analytical procedures

3.1.1 Ammonia

Ammonia was determined according to DEV (1983) procedure E5 (DIN 38406). The two reagents used for analysis of ammonia were:

Reagent A; 13 g Natriumsalicylate, 13 g Tri-Natriumcitrate-Dihydrat and 0.097 g 2-Nitroprussidnatrium-Dihydrat were dissolved in 500 ml deionized H_2O. Reagent B; 1.6 g NaOH and 0.1 g Dichlorocyanuric acid-Na-Dihydrat were dissolved in 50 ml deionized H_2O.

Ammonia ions reacts at a pH value of about 2.6 with hypochlorite and sodium salicylate in the presence of Sodium Pentacyanonitrosylferrate as a catalyst to a blue-colored product which could be measured in a spectrophotometer at 655 nm.

For analysis, 0.125 ml reagents A and 0.125 ml reagent B were added into 1 ml sample, and then measured after 1–3 hours incubation at room temperature.

3.1.2 Nitrite

Nitrite was determined colorimetrically according to DEV (1983) procedure D28 (DIN 38405). The reagent used for the method is made by dissolving 20 g Sulfanilamide, 1 g N-(1-Napthyl)-ethylendiamine-dihydrochloride and 50 ml O-Phosphoric acid (1.71 g mL^{-1}) in 500 ml deionized H_2O.

The reaction of nitrite with sulfanilamide and N-(1-Naphthyl)-ethylene-diamine dihydrochloride in the presence of phosphoric acid results in a pink-colored diazo dye, which can be measured in a spectrophotometer at 540 nm.

For analysis, 0.020 ml reagent is added into 1 ml sample and then measured after 20–30 minute incubation at room temperature.

3.1.3 Nitrate

Nitrate was determined colorimetrically according to standard methods APHA (1995). No reagent is needed in this method. NO_3^- ions in samples absorb light at 220 nm. Nitrite ions interference with this method, if the nitrite concentration in samples was higher than 0.65 mg NO_2^--N L^{-1}, Amido-sulfonic acid must be added into samples.

3.1.4 Alkalinity

The alkalinity of the samples was measured according to APHA (1995). The sample alkalinity were calculated after titration to a pH value of 4.5 with 0.02 M HCl and expressed as $CaCO_3$ mg L^{-1}.

3.1.5 Total solids, ash content (minerals)

The solids content was determined by using standard methods (APHA 1995). Due to the high salt contents, samples were centrifuged at 10000 rpm for 10 minute and the pellet was re-suspended twice with the same volume of deionized water. The samples were evaporated in a weighed vessel and dried to constant weight. For determining of the total solids (TS), the sample was dried at 103 to 105°C. The ash content was determined after sample incubation at 600°C for 6 h. The organic fraction of TS (OTS or volatile solids) was obtained by subtraction of the mineral content (residual ash after oxidation at 600°C) from the total solids.

3.1.6 Dissolved oxygen (DO), pH, salinity

The DO, pH, temperature, and electrical conductivity (salinity) were measured with respective standard probes attached to a multimeter (Inolab multi level 1, WTW Weilheim, Germany). The salinity unit used in the report is percent (%) referring to microSiemens cm^{-1}.

3.1.7 Microscopical examination

Microscopic observations of the reactor contents were carried out regularly, using phase contrast microscopy (Zeiss Standard 14) at 1000 x magnification.

3.1.8 Chemicals

All chemicals used were of analytical grade and were purchased from VWR/Merck (Darmstadt), Fluka (Taufkirchen), or Roth (Karlsruhe, Germany)

3.2Basic calculations

3.2.1 Hydraulic retention time

The hydraulic retention time (HRT) is the average retention time of wastewater in the reactor.

It was calculated as the ratio of liquid volume (V liquid) in the reactor and the flow rate (Q)

Eq. 3.1: $$HRT=\frac{V_r}{Q} \quad \left[\frac{L}{L.d^{-1}}=day\right]$$

HRT : hydraulic retention time (d)
V_r : liquid volume of the reactor (L)
Q : flow rate ($L\ d^{-1}$)

3.2.2 Ammonia loading rate

The volumetric ammonia loading rate was the amount of ammonia applied to the reactor volume per day. It was calculated according to Equation 3.2

Eq. 3.2: $$ALR=\frac{C_{inf}[NH_4^+] \times Q}{V_r} \quad \left[\frac{mg.L^{-1} \times L.d^{-1}}{L}=mg.L^{-1}.d^{-1}\right]$$

ALR : volumetric ammonia loading rate ($mg\ L^{-1}\ d^{-1}$)
C_{inf} : concentration of the ammonia ($mg\ L^{-1}$)
Q : flow rate ($L\ d^{-1}$)
V_r : liquid volume of the reactor (L)

3.2.3 External recirculation rate

The amount of wastewater from the overflow of reactor that was pumped back to the inlet pipe (L h^{-1}).

$$Eq.\ 3.3: \quad Q_r = \frac{V_m}{T} \left[\frac{L}{h} = L.h^{-1}\right]$$

Q_r : recirculation rate (L h^{-1})
V_m : volume medium pumped back to the inlet pipe (L)
T : time needed for a given volume of medium to be pumped back

3.2.4 Specific surface area and porosity of carrying materials

Porosity was simply determined by first filling a glass with water. Then, the supporting materials, which their volumes (V_{sm}) are already known, were filled into the glass. The water was poured out and was collected to obtain the water volume (V_w). Porosity in percentage was obtained from V_w/V_{sm}.

The specific surface area was determined by calculating surface area per volume from the used supporting materials.

3.2.5 Aeration rate

The gassing rate (aeration rate) is the amount of gas/air (liter) applied to a reactor per hour.

$$Eq.3.4: \quad Q_g = \frac{V_g}{T} \left[\frac{L}{h} = L.h^{-1}\right]$$

Q_g : gas rate (L h^{-1})
V_g : amount gas supplied into a reactor (L)
T : time needed for a given amount of gas supply into a reactor (h)

3.3 Experimental design

3.3.1 Continuously operated reactors

3.3.1.1 Fixed-bed reactors

Two cylindrical fixed-bed reactors (FBR A and B) with 9 cm internal diameter and 45 cm height were used (see Figure 3.1).

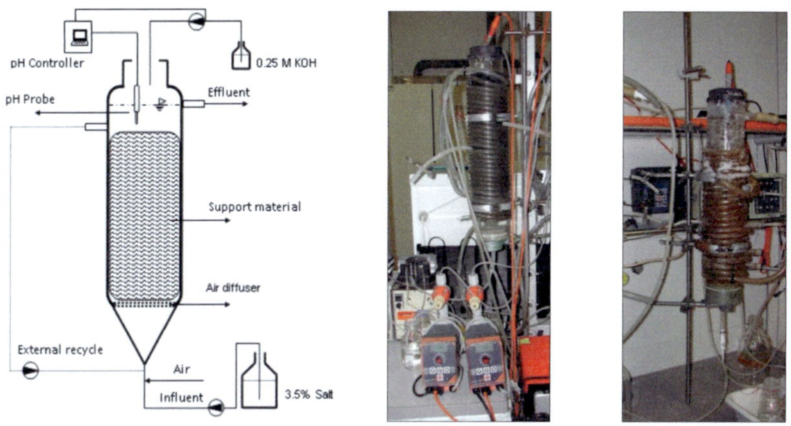

(a) (b) (c)

Figure 3.1 Scheme (a) and photos of the fixed-bed reactors (b. FBR A, c. FBR B) used for nitrification experiments

The feed was pumped into the reactor at the bottom with a Gilson Minipuls 3 pump (Abimed, Dreieich, Germany), and effluent was flowing out at the top of the reactor through an overflow pipe, maintaining a working volume of 2.1 L. Polyethylene/clay sinter lamellas (Figure 3.2a) PELIA, Herding, Amberg, Germany with a specific surface area of 440 m^2 m^{-3}, a fixed-bed area of 0.46 m^2, and a porosity of 59% were used as support material in FBR A and porous ceramic rings (Figure 3.2b; Poro Ring, Tropical) with a specific surface area of 934 m^2 m^{-3}, a fixed-bed area of 0.60 m^2, and a porosity of 38% in FBR B.

Both FBRs were filled with seawater from Hafen Büsum that contained 100 mg N L^{-1} (380 mg L^{-1} NH_4Cl). During the first 40 days, the FBRs were operated batch wise until all ammonia and nitrite was oxidized to nitrate. Afterwards, the FBRs were operated continuously at different operational conditions (Table 3.1). The ammonia loading rate (ALR) was increased by reducing the hydraulic retention time (HRT) or by increasing the ammonia concentration. The reactors were aerated from the bottom with an air diffuser at a flow rate of 2.5 L h^{-1}, which was adjusted by a micro-valve and controlled by a TG 05 gas meter (Ritter, Bochum-Langendreer, Germany). This maintained an oxygen concentration of ≥5 mg L^{-1} and supported a homogeneous distribution of the feeding. To further improve liquid mixing, an external recirculation loop was installed from the overflow to the inlet pipe for both FBRs.

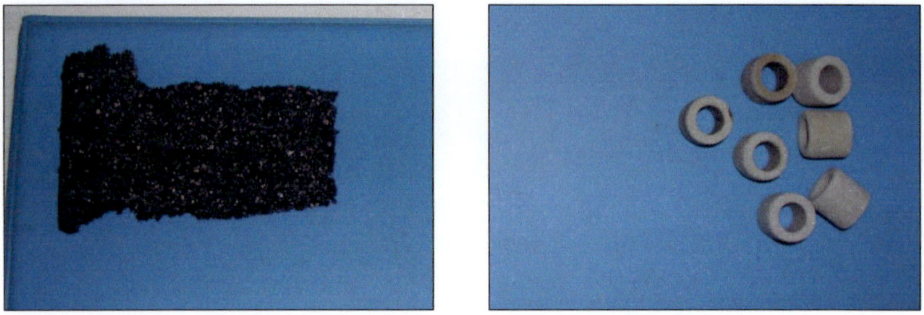

(a) (b)

Figure 3.2 Support materials used for nitrification experiments (a) Polyethylene/clay sinter lamellas, (b) porous ceramic rings

The reactor effluent was re-circulated at a flow rate of 8 L h^{-1} into the inlet pipe of the reactor with a Watson-Marlow pump (model 604 U, Falmouth, Cornwall, England). The pH was kept at 7.6 by a pH titrator (Dulcometer D1C, Prominent, Heidelberg, Germany) and a dosing pump (Prominent Gamma 74) with 0.25 M KOH solution. The reactors were run at a room temperature.

Table 3.1 Operational phases a–d of FBR A and FBR B

Phase	a	b	c	d		
Days	0–55	56–155	156–186	187–194	195–215	216–305
HRT (days)	1.25	1	1	1	1	1
pH adjustment with	KOH	KOH	KOH	$NaHCO_3$	KOH	KOH
NH_4^+-N_{in} (mg L^{-1})	104	104	130	130	130	130
ALR mg NH_4^+-N L^{-1} day^{-1}	83	104	130	130	130	130
External recirculation	–	–	–	–	–	8 L h^{-1}

3.3.1.2 Ceramic rings fixed bed reactors under salinity fluctuation, started with 3.5% salt

Three Plexiglas cylindrical fixed–bed reactors (FBR C, D and E) with a working volume of 0.2 L were run in parallel (Figure 3.3). The fixed-bed consisted of 14 pieces of porous ceramic rings taken from FBR B which was running continuously at an ammonia loading rate (ALR) of 130 mg NH_4^+-N L^{-1} day^{-1} and a HRT of 1 day under steady state conditions. The nitrifying biomass-containing porous ceramic rings were filled into the upper compartment of FBR C, D and E. Synthetic seawater containing 60 mg $NH4^+$-N L^{-1} and a salt concentration of 3.5% (w/v) was pumped into the lower compartment of the reactor with a Gilson Minipuls 3 pump (Abimed, Dreieich, Germany) at a flow rate of 0.2 L day^{-1} to maintain a HRT of 1 day.

Sodium bicarbonate was applied in synthetic medium to keep the pH at 8$\pm$0.2 with a fixed ratio of alkalinity:ammonia nitrogen (g $CaCO_3$/g NH_4^+-N of 7.1: 1). To keep the dissolved oxygen concentration (DO) above 5 mg L^{-1}, tube aeration with a flow rate of 2 L h^{-1} was applied at the lower compartment, adjusted by a NUPRO micro valve and controlled with a TG 05 gas meter (Ritter, Bochum, Germany).

(a) (b)

Figure 3.3 Scheme (a) and photo (b) of the fixed-bed reactors – FBR C, D and E

During experimentation salinity in FBR C, D and E were stepwise decreased or increased every 14 days as shown in Table 3.2. Six phases were distinguished in each FBR with fixed salt concentrations. The FBRs contained 3.5% of salinity in the first phase to compare nitrification activity of the inocula that were taken from the supply reactor. Then, the salt concentration of the medium was decreased to minimally 0.034 in two steps via 0.5 (FBR C), 1 (FBR D) or 2% (FBR E) or increased in two steps via 3.5% to 5, 7, or 9%, respectively and finally adjusted to 3.5% salinity in all reactors.

Table 3.2 Salt concentration in percent during operational phases I–VII of FBR C, D and E

Phase – days	I (0–15)	II (16–27)	III (28–43)	IV (44–57)	V (58–74)	VI (75–85)	VII (86–96)
FBR C	3.5	0.5	0.034	3.5	9	3.5	3.5
FBR D	3.5	1	0.034	3.5	7	3.5	3.5
FBR E	3.5	2	0.034	3.5	5	3.5	3.5

An adaptation period of about two weeks was allowed for each change of the salt concentration. The changes in the salt concentration of the reactor influent were induced by changing the influent concentration to the desired new concentration. The time needed for salinity changes up or down to the new desired concentration was calculated by Equation 3.5.

Eq.3.5: $$t = \frac{\ln\left[\frac{C_t}{C_0}\right]}{\pm k}$$

T= time (hours), C_t= targeted salinity (%), C_0= initial salinity (%)
k = rate of concentration change (h^{-1}), positive value when $C_t > C_0$

3.3.1.3 Behavior of polyethylene/clay sinter lamellas containing fixed bed reactors under salinity fluctuation, started with 3.5% NaCl

Four Plexiglas cylindrical fixed–bed reactors (FBR F, G, H and I) with a working volume of 0.2 L were continuously run in parallel (Figure 3.4). Polyethylene/clay sinter lamellas taken from FBR A at steady state conditions were cut into pieces (length = 5 cm; width = 1.5 cm; thick = 0.4 cm), of which 9 were used as fixed bed material in the FBRs. The pieces of Polyethylene/clay sinter lamellas were arranged in reactors in a way, so that there was a free room at the bottom for mixing with a magnetic stirrer. Dissolved Oxygen concentration measurements and pH adjustments in the reactors were conducted as in FBR C, D or E (see section 3.3.1.2).

Phases of decreasing or increasing salinity during continuous operation of FBR F, G, H and I are shown in Table 3.3. In FBR L (a control reactor) salinity was kept at 3.5% in all phases except for phase IV.

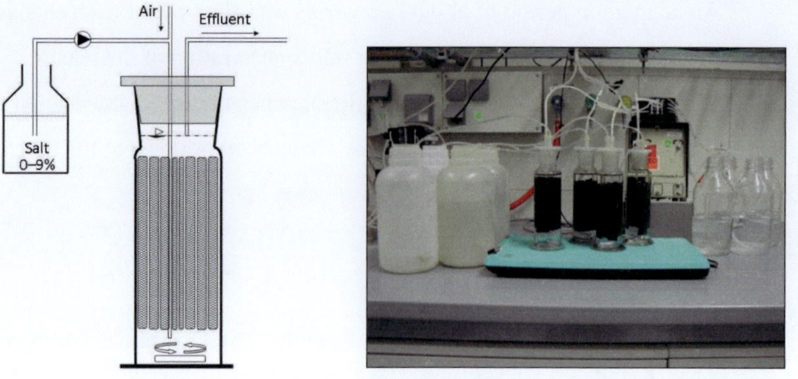

(a) (b)

Figure 3.4 Scheme (a) and photo (b) of the fixed-bed reactors – FBR F, G, H and I

Reactors were operated at an ammonia loading rate (ALR) of 130 mg NH_4^+-NL^{-1} day^{-1} and at 1 d HRT. Synthetic seawater containing 100 mg NH_4^+-N L^{-1} and a salt concentration of 3.5% (w/v) was pumped into the lower compartment of the reactor with a Gilson Minipuls 3 pump (Abimed, Dreieich, Germany) at a flow rate of 0.2 L day^{-1} to maintain a HRT of 1 day

Table 3.3 Salt concentration in percent during operational phases I–VIII of FBR F, G, H and I

Phase – days	I (0–6)	II (7–16)	III (17–35)	IV (36–49)	V (50–63)	VI (64–80)	VII (81–90)	VIII (91–99)
FBR F	3.5	3.5	0.5	0.034	3.5	9	3.5	3.5
FBR G	3.5	3.5	1	0.034	3.5	7	3.5	3.5
FBR H	3.5	3.5	2	0.034	3.5	5	3.5	3.5
FBR I	3.5	3.5	3.5	0.034	3.5	3.5	3.5	3.5

3.3.1.4 Behavior of polyethylene/clay sinter lamellas containing fixed-bed reactors under salinity fluctuation with a starting concentration of 0.03% salt

Five Plexiglas cylindrical fixed–bed reactors (FBR J, K, L, M and N) with a working volume of 0.2 L were run in parallel (Figure 3.5). The fixed-bed consisted of 9 pieces of fresh Polyethylene/clay sinter lamellas (length = 5 cm; width = 1.5 cm; thick = 0.4 cm) which were arranged in the reactors to leave a free volume at the bottom for magnetical stirring.

As inocula, water samples (each 150 ml) taken from a "Brackwasser"-location at the North Sea and already activated for nitrification were filled into the reactors. Before activation by ammonia addition, the water had a pH of 7.1, an alkalinity of 65 mg L^{-1} (as $CaCO_3$) and a salinity of 0.05%. The concentration of ammonia, nitrite and nitrate was 0.5, 0.1 and 10.5 mg-N L^{-1}, respectively.

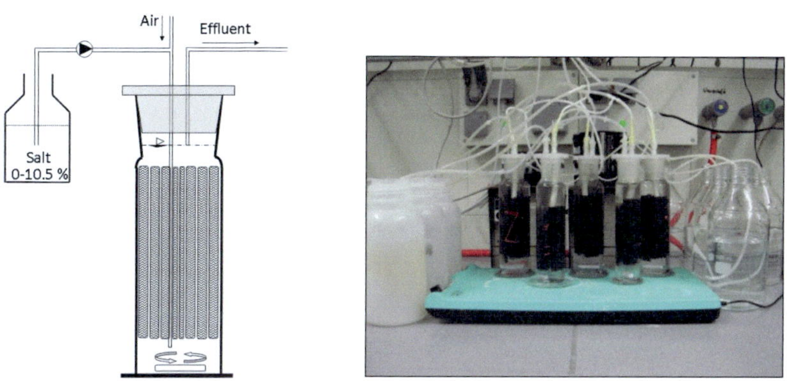

(a) (b)

Figure 3.5 Scheme (a) and photo (b) of the fixed-bed reactors – FBR J, K, L, M and N

The FBRs were operated batch wise for two days, so microorganisms had time to attach onto the supporting materials. During the batch wise operation aeration at a flow rate of 2 L h^{-1}) and mixing with a magnetic stirrer was already maintained.

For continuous operation, synthetic seawater containing 120 mg NH_4^+-N L^{-1} was pumped into the lower compartment of the reactor with a Gilson Minipuls 3 pump (Abimed, Dreieich, Germany) at a flow rate of 0.2 L day^{-1} to maintain a HRT of 1 day. Forced salt fluctuations during continuous operation are shown in Table 3.4. Reactors were started with 0.03% NaCl, which was the salinity of the inocula. Adjustment for pH and dissolved oxygen in the reactors was done as in FBR C, D an

Table 3.4 Salt concentration in percent during operational phases I–VII of FBR J, K, L, M and N

Phase – days	I (0–8)	II (9–24)	III (25–36)	IV (37–52)	V (53–70)	VI (71–89)	VII (90–99)
FBR J	0.034	0.034	0.5	0.034	10.5	0.034	0.034
FBR K	0.034	0.034	1	0.034	9	0.034	0.034
FBR L	0.034	0.034	2	0.034	7	0.034	0.034
FBR M	0.034	0.034	3.5	0.034	5	0.034	0.034
FBR N	0.034	0.034	0.034	0.034	0.034	0.034	0.034

3.3.1.5 Operation of fixed-bed reactors with polyethylene/clay sinter lamellas under gradual and abrupt ammonia loading rates

Three Plexiglas cylindrical fixed-bed reactors (Figure 3.6; FBR O, P and Q), similar to FBR C, D and E (section 3.3.1.2) were assembled using fresh polyethylene/clay sinter lamellas as carrier material

Figure 3.6 Photo of the fixed-bed reactors – FBR O, P and Q

As inoculum, a nitrifying biofilm from inside the silicon tube of the external water recirculation system of FBR A was removed and added as a suspension. The reactor was filled with synthetic seawater (3.5% of salinity) and was operated batch wise for 20 days to acclimatize and immobilize the biomass on the supporting material. During the acclimatization, the reactors were fed twice with ammonia solution to reach 100 mg L^{-1} NH_4^+-N (t = 0 and t = 10 day).

After 20 days reactors were continuously operated as indicated in Table 3.5. Several phases were distinguished, concerning the ammonia concentration in the inlet. Reactor O is a control reactor, in which the ammonia concentration was kept constant during the whole experiment. Reactor P represents a steady increase of the ammonia concentration and reactor Q represents a sudden increase of ammonia. Ratio of Bicarbonate as $CaCO_2^-$ ammonia as NH_4^+-N in synthetic seawater was 7.1: 1 to maintain a pH of 8 ± 0.2.

A HRT of 1 day was achieved with controlling inlet pump (Gilson Minipuls 3 pump – Abimed Dreieich, Germany) at rate of 0.2 L d^{-1}. Applied pipes for inlet, outlet and aeration in the reactors were the same with that in FBR C, D, and E (section 3.3.1.2).

Table 3.5 Ammonia concentration in percent during operational phases I–V of FBR O, P and Q

Phase – days	I (0–20)	II (21–72)	III (73–89)	IV (90–105)	V (106–140)
HRT (days)	Batch	0.5	0.5	0.5	0.5
NH_4^+-N_{in} (mgL^{-1})					
– FBR O		40	40	40	40
– FBR P		40	85	125	250
– FBR Q		40	250	250	250

3.3.1.6 Ammonia oxidation rates (AORs) and nitrite oxidation rates (NORs)

For determining the AORs and NORs of FBR A and B at the respective days and for FBR C- Q at the end of every phase (Table 3.2; 3.3; 3.4; 3.5), the reactors were operated in batch mode for two days.

When all residual ammonia or nitrite from continuous operation was used up in the batch-mode, respective nitrogen sources for AOB (NH_4Cl, 60 mg N L^{-1} for FBR A and B, 30 mg N L^{-1} for FBR C–Q) or NOB ($NaNO_2$, 30 mg N L^{-1} for FBR A and B, 15 mg N L^{-1} for FBR C–Q) for nitritation and nitratation were added. AORs and NORs in the three FBRs were determined from the slope of ammonia or nitrite concentrations. For AOR measurements, $NaHCO_3$ solution was added to keep the pH at 8.

When all ammonia and nitrite was used up after AOR and NOR determinations, continuous operation was proceed by replenishing the ammonia concentration to those values before the batch mode.

3.3.2 Batch Experiment

3.3.2.1 Batch reactors for selection of biomass

Water and sediment sampling: Water and sediment samples a–d were collected from different costal locations at the North Sea in January 2008. The characterization of the samples is shown in Table 3.6. The water samples (10 L) and the sediment samples (5 L) were filled into sterilized canisters and stored at 4°C in a cool box during transportation to the laboratory.

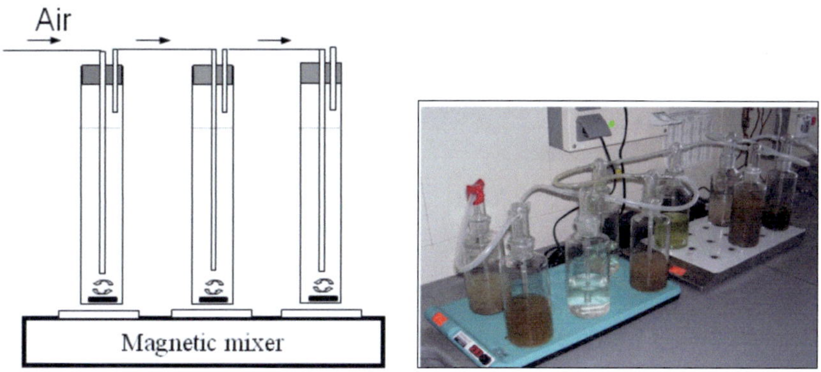

(a) (b)

Figure 3.7 Scheme (a) and photo (b) of batch reactors for biomass selection

For determination of the nitrifying activity in batch assays, water or sediment samples (250 mL each) from the different collection places at the North Sea were filled into glass reactors of 500 mL total volume. A magnetic stirrer was used to mix the samples in the reactors. The reactors were aerated with 2 L h^{-1} air blown in via a syringe needle just above the stirrer to keep the DO above 5 mg L^{-1}. After addition of 50 mg N L^{-1} (190 mg L^{-1} NH_4Cl), incubation at room temperature (20–23°C) was started. At different time intervals, liquid samples were taken for analysis of dissolved oxygen (DO), pH, ammonia, nitrite, and nitrate. For pH correction, 1–3 mL 0.25-M KOH was added to the reactors until a pH >8.0 was reached. The DO in all samples was kept above 5 mg L^{-1}.

Table 3.6 Salinity and alkalinity of the used seawater/mud samples

Sample		Conductivity ($mScm^{-1}$)	Alkalinity (mgL^{-1}) as $CaCO_3$
a.	Seawater from Hafen Büsum	31.9	120
b.	Seawater/mud mixture from town Norden	37.5	400
c.	Brine water from town St. Peter-Ording	1.3	53
d.	Seawater from town St.Peter-Ording	32.1	130

3.3.2.2 Biomass growth in porous ceramic rings

For this experiment, fresh porous ceramic rings (36 pieces) were put into the reactor (FBR B), just above the old fixed bed. The porous ceramic ring fixed bed reactor (FBR B) that was operated continuously for more than 500 day was continued to be operated at a HRT of 1 day, a DO of above 5 mg L^{-1} and pH of 8. Ammonia concentration in the feeding was 50 mg NH_4^+-N L^{-1}. Neither ammonia nor nitrite was detected in the outlet under the operation conditions.

At days 1, 5, 12, 20, 33 and 50, six porous ceramic rings were taken out and then ammonia and nitrite removal rate by biofilm in the ceramic rings was separately measured. The measurement of rates was done in small cylindrical glass reactors with initial concentration of ammonia and nitrite was 10 mg N L^{-1}.

The small cylindrical glass reactors have an internal diameter of 3 cm, 15 cm height and 50 ml working volume. The reactors were aerated from the bottom with 2 L^{-1} air. Because these reactors were also used in other assays under different conditions, they were called “small reactors” if referring to biomass balancing (Figure 3.8)

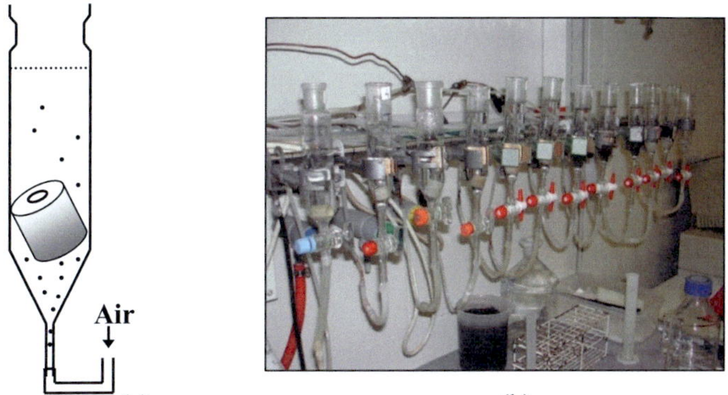

(a) (b)

Figure 3.8 Scheme (a) and photo (b) of small cylinder glass reactors for batch assays

Triplicate measurements for both rates (nitritation, nitratation) were conducted to obtain representative results.

3.3.2.3 *Ammonia and nitrite oxidation rates of the biofilm on porous ceramic rings taken from the same level of fixed-bed reactors*

Porous ceramic rings covered with nitrifying biomass, were taken from FBR B, drained for 30 minutes and then filled into the small reactors. Synthetic seawater (3.5% salinity) with an ammonia concentration of 10 mg NH_4^+-NL^{-1} for measurement of the ammonia oxidation rate and a nitrite concentration of 10 mg NO_2^- -N L^{-1} for measurement of the nitrite oxidation rate was added. The pH was kept at 8 by adding $NaHCO_3$ solution.

Ammonia and nitrite concentrations in the reactors were measured spectrophotometrically every 1.5 hour.

3.3.2.4 *The effect of sodium azide and allylthiourea (ATU) on ammonia and nitrite oxidation rates in biofilm nitrifiers*

Porous ceramic rings were taken from FBR B, drained for 30 minutes and filled into the small reactors. To determine the effect of sodium azide and ATU on ammonia or nitrite oxidation rates, measurement of the rates was conducted twice. In the first measurement 50 ml synthetic wastewater (3.5% salt) with 10 mg NH_4^+- N L^{-1} (for AOR) and 10 mg NO_2^--NL^{-1} was added to the small reactors. Decreasing ammonia or nitrite concentration was monitored to determine AORs and NORs, respectively.

Table 3.7 The conditions of assays during batch experiments

Factor	Ammonia oxidation rate			Nitrite oxidation rate		
	Measurement			Measurement		
	1st	2nd	3rd	1st	2nd	3rd
Sodium Azide (µg L^{-1})	0	0.1;0.3:0.5; 20;40;80		0	0.1;0.3:0.5; 20;40;80	
Allythiourea (mg L^{-1})	0	0.02;0.043:0.1; 0.2;0.4;1;4		0	0.02;0.043:0.1; 0.2;0.4;1;4	
Ammonia (mg N L^{-1})	0.01	0.01;0.25;0.5;1;2.3.5;5	0.01	0	0;0.25;0.5;1;2;3.5;5	0
Nitrite (mg N L^{-1})	0	0;0.02;0.04;0.075; 0.125;0.25;0.4	0	0.01	0.01;0.5;1; 1.5; 3.5;5;10	0.01
Nitrate (g N L^{-1})	0	0;0.5;1;1.5; 3.5;10	0	0	0;0.5;1;1.5; 3.5;10	0
Temperature (°C)	22.5	6;12,5:27.5;32.5;40	22,5	22,5	6;12.5:27.5;32.5;40	22.5
Salinity (%)	3.5	0.03; 0.5; 1; 2; 3.5; 5; 7; 9	3.5	3.5	0.03; 0.5; 1; 2; 3.5; 5; 7; 9	3.5

For the second measurement, the medium in the reactors was drained slowly to avoid detachment of the biofilm and then the new medium containing 3.5% salt and 10 mg NH_4^+-N or 10 mg NO_2^--N as well as appreciated concentration of ATU or sodium azide was added into the reactors (Table 3.7). The AORs and NORs were obtained by monitoring decreasing ammonia or nitrite concentration and were then compared with AOR and NOR of first measurement to determining effects of ATU and sodium azide.

3.3.2.5 *The effect of NH_4^+ and NO_2^- on ammonia/nitrite oxidation rate by biofilm nitrifiers*

Porous ceramic rings were taken from FBR B, drained for 30 minutes and filled into the small reactors.

- First measurement

For measurement of the basic AOR or NOR, 50 ml of synthetic medium (3.5% salt) with 10 mg NH_4^+-N L^{-1} or 10 mg NO_2^--N L^{-1} was added to the reactors. The AORs and NORs were determined by monitoring decreasing ammonia or nitrite concentration respectively.

- Second measurement

i. Effect of ammonia on AOR

Synthetic medium (50 ml) with desired concentration of ammonia was added into appropriated reactors (Table 3.7).

ii. Effect of ammonia on NOR

Synthetic medium (50 ml) with 10 mg NO_2^--N L^{-1} and desired concentration of ammonia was added into appropriated reactors.

iii. Effect of nitrite on AOR

Synthetic medium (50 ml) with 10 mg NH_4^+-N L^{-1} and desired concentration of nitrite was added to the reactors.

iv. Effect of nitrite on AOR

Synthetic medium (50 ml) with desired concentration of nitrite was added into appropriated reactors.

v. Effect of nitrate on AOR and NOR

Synthetic medium (50 ml) with 10 mg NH_4^+-N L^{-1} or 10 mg NO_2^- -N L^{-1} and desired concentration of nitrate was added.

- Third measurement

If necessary, a third measurement for determining recovery ability was conducted by draining the medium and changing the medium like for the first measurement.

3.3.2.6 *The effect of temperature on ammonia/nitrite oxidation rate by biofilm nitrifiers*

Three series of this experiment were conducted. After draining for about 30 minutes, porous ceramic rings from FBR B were put into small reactors and then 50 ml of synthetic medium (3.5% salt) with initial ammonia and nitrite concentration of 10 mg N L^{-1} was filled in. The first series of experiments was conducted at room temperature (22.5°C). The second series of experiments were conducted at different temperatures (Table 3.7) by placing the reactors into a thermostated incubator. After a slow drainage the new medium (3.5% salt, 10 mg N L^{-1} of ammonia or nitrite) was added into the reactors. The third series were conducted at 22.5°C to determine the recovery ability of the biofilm.

Ammonia and nitrite oxidation rates were quantified for each serie. Determination of AOR and NOR was based by decreasing of ammonia or nitrite concentration.

3.3.2.7 *The effect of salinity on ammonia/nitrite oxidation rates of immobilized nitrifiers*

Porous ceramic rings from FBR B were drained for 30 minutes and filled into small reactors.

- First measurement

The basic AORs or NORs were measured by adding synthetic medium (3.5%, 10 mg NH_4^+-N or 10 mg NO_2^--N) into the reactors. Decreasing ammonia or nitrite concentration represented the AORs and NORs.

- Second measurement

After the medium was drained, 50 ml of synthetic medium (10 mg NH_4^+-N or 10 mg NO_2^--N) with desired salinity, as shown in Table 3.7, was added into the reactors. The AORs and NORs were determined based on decreasing ammonia and nitrite concentration.

- Third measurement

The medium was again drained and the same synthetic medium as it for the first measurement was added.

3.3.2.8 The effect of salinity on ammonia/nitrite oxidation rate by suspended nitrifiers

For testing oxidation rates of suspended nitrifiers, two reactors for enrichment of suspended nitrifying bacteria were assembled, the first as a continuous flow stirred tank reactor (CSTR) and the second as a sequencing batch reactor (SBR). The CSTR consisted of a 2-liter glass flask equipped with a magnetic stirrer and aerated from the bottom of the reactor with air (Figure 3.9).

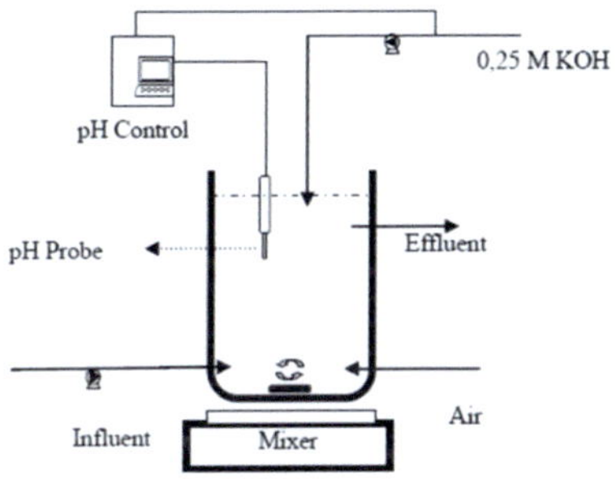

(a)

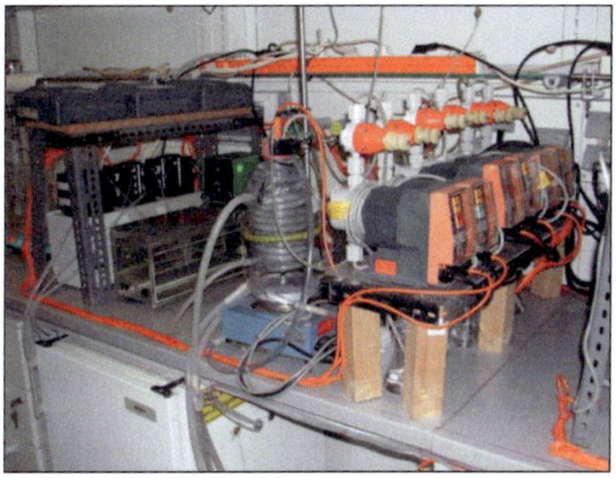

(b)

Figure 3.9 Scheme (a) and photo (b) of the CSTR for enrichment of nitrifying bacteria.

Synthetic wastewater containing 3.5% salinity and 104 mg NH_4^+-N L^{-1} was pumped into the reactor at a flow rate of 1 L d^{-1} to maintain a HRT of 2 days. The reactor was operated at temperature of 22.5°C and a pH of 8 ±02 which was adjusted with 0.25 M KOH solution by a pH titrator (Dulcometer D1C, Prominent, Heidelberg, Germany) and a dosing pump (Prominent Gamma 74). The effluent of the reactors was collected in a glass flask, which allowed for biomass settling. Supernatant was then drained off and biomass was added back to the reactors.

The SBRs consisted of 1-liter glass flasks equipped with a magnetic stirrer and aerated with a diffuser. A fresh water-sediment mixture (1000 ml), taken from a place near by Büsum, was filled into the reactors. The SBRs were also operated at pH 8±0.2 using sodium carbonate as a buffer and were incubated at a temperature of 22.5 ^{0}C. During feeding periods, ammonia solution to a final concentration of 104 mg NH_4^+-N L^{-1} was added into the reactors. When ammonia and nitrite were completely oxidized to nitrate, a second feeding was conducted. After three-feeding periods, the reactor was allowed to stand for 3 hours with no mixing and aeration for biomass settling.

About 800 ml volume of supernatant was carefully drained from the reactor. Mixing, aeration and feeding were started immediately after adding 800 ml of synthetic wastewater containing 0.03% of salt as salinity of the inoculum.

Biomass from above mentioned CSTRs and SBRs was used for determining the effect of salinity on ammonia and nitrite oxidation rates by suspended nitrifiers

Samples were taken from the CSTR and the SBR (each 50 ml). After centrifugation at 10000 rpm for 5 minutes, about 90% of supernatant was drained. Then, the pellet was re-suspended in the rest of the supernatant and divided homogeneously into 100 ml-glasses. The glasses were filled by 50 ml of synthetic wastewater containing 60 mg NH_4^+-N L^{-1} or 25 mg NO_2^- -N L^{-1} with different salinity (0.03; 0.5; 1; 2; 3.5; 5; 7; 9%).

After aeration and mixing was set-up, AORs or NORs were determined by monitoring decreasing of ammonia or nitrite concentration.

To examine the effect of salinity changes, a second measurement of AORs and NORs was done. The medium in the glasses (50 ml) was centrifuged at 10.000 rpm for 5 minute; the pellet was kept back in the glasses. Synthetic medium (3.5% or 0.03% of salinity) with the same concentration of ammonia or nitrite as before was added. AORs or NORs were determined after aeration and mixing was set-up.

3.3.2.9The effect of aerobic and anaerobic storage of enrichment cultures on the ammonia oxidation rate

The effect of aerobic-anaerobic storage of enrichment cultures on the ammonia oxidation rate was investigated using suspended bacteria as inocula. The suspended bacteria were taken from the CSTR nitrifying reactor.

To determine the effect of aerobic storage on AOR or NOR 25 ml of biomass taken from the CSTR reactor was filled into 100 ml-serum flasks. The initial pH was 8 ± 0.2 by adding sodium bicarbonate.

- Aerobic conditions

The nitrifying bacteria in the flasks were mixed with a magnetic stirrer. The mixing rate was adjusted at 300 rpm to obtain homogeneously suspended biomass and to supply enough oxygen. To reduce evaporation of medium the flask was closed by porous silicon stoppers. Decreasing ammonia or nitrite concentrations were monitored and the respective AORs and NORs were determined.

- Anaerobic condition

The flasks were closed with butyl rubber stoppers and aluminum caps. The initial pH of all flasks was 8.0. To obtain anaerobic conditions, the headspace of the flasks was flushed three times with nitrogen gas by applying a pressure of 1 bar and then releasing the pressure in the headspace with a syringe to maintain an atmosphere pressure. The nitrifiying culture in the flasks was mixed with a magnetic stirrer.

Chapter 4

RESULTS

4.1 Characterization of marine water and coastal sediment samples as natural sources of halophilic nitrifier

Sludge or water samples originating from marine environments usually contain microorganisms that have the ability to be metabolically active under saline conditions. Thus, seawater and sediment samples from different coastal regions at the North Sea were collected and incubated under selective conditions to enrich nitrifiers. The characterization of the samples for parameters such as salinity, alkalinity, ammonia, nitrite and nitrate concentrations is summarized in Table 4.1.

Table 4.1 Characterization of the used seawater/mud samples

Sample	Conductivity ($mScm^{-1}$)	Alkalinity (mgL^{-1}) as $CaCO_3$	NH_4^+-N (mgL^{-1})	NO_2^--N (mgL^{-1})	NO_3^--N (mgL^{-1})
A. Seawater from Hafen Büsum	31.9	120	0	0.11	2.3
B. Seawater/mud mixture from the town Norden	37.5	400	0.2	0.32	10.3
C. Brine water from the town St. Peter-Ording	1.3	53	0.9	0.37	5.1
D. Seawater from the town St. Peter-Ording	32.1	130	0.6	0.68	1.7

It can be seen that the ammonia concentration in all four samples was less than 1 mg $L^{-1.}$ In seawater/mud sample B from a “brine water pond” that remained in the mud during low tide and that was fed with treated sewage effluent at the harbor of the city of Norden, the nitrate concentration was the highest with 10.3 mg N L^{-1} (Table 4.1), most probably due to the ongoing nitrified domestic wastewater supply during low tide. Sample B revealed the highest salinity and the highest alkalinity, presumably caused by its undiluted seawater and its high mud content. The ammonia concentration of sample C was much higher than that of sample D, due to the pollution with domestic wastewater. Wastewater discharged at the coast into brine water (sample C) caused an elevated content of ammonia and nitrate. At the point where the brine water flows to the open sea (sample D), ammonia and nitrate concentrations were much lower due to dilution with seawater.

The nitrifying capability of seawater/mud samples was tested by supplementing 50 mg N L^{-1} ammonia. Portions of 250 mL were filled into batch reactors for nitrification by the autochthonic bacteria. Ammonia depletion, nitrite as well as nitrate formation and utilization were analyzed. Figure 4.1 shows that ammonia oxidation began in all samples after a lag phase of approximately 10 days, and led to the formation of abundantly nitrite (samples A, B, and D) or nitrate (sample C).

Insufficient alkalinity in samples A, C, and D resulted in incomplete ammonia conversion to nitrite or nitrate and a pH drop below or far below 7. Ammonia was, however, completely converted to nitrite and then to nitrate in seawater/mud sample B, which had a high alkalinity and a stable pH with a minimal value of 7.5 (Fig. 4.1b). When the second portion of ammonia was fed to this reactor at day 42, ammonia was oxidized rapidly to nitrite. The nitrite was further oxidized to nitrate within less than 10 days, indicating the successful enrichment of halophilic or at least halotolerant nitrifiers.

In the reactor with brine water (sample C), very little ammonia was oxidized at the beginning due to the rapid drop of the pH (sample C had the lowest alkalinity of all samples), but nitritation and nitration proceeded shortly after titrating the pH to 8.7 (Fig. 4.1c, day 42). The strict dependence of ammonia oxidation on an alkaline pH can be seen in the three reactors where seawater samples with a low alkalinity were incubated for the establishment of halophilic nitrification (Fig. 4.1a, c, d). In these reactors, the pH dropped rapidly when ammonia oxidation started and nitrite was produced. When the pH was raised to around 8.5 by addition of KOH at days 42 and 52, ammonia oxidation was completed and nitrite conversion to nitrate started (Fig. 4.a, c) or was finished (Fig. 4.1d).

The initial AOR of the enrichment culture obtained with seawater from Hafen Büsum (sample A), after the lag phase and after pH correction was around 11.5 mg N L^{-1} day^{-1} and the enrichment from brine and seawater of St. Peter Ording (samples C and D) were 7.3 and 6.6 mg N L^{-1} day^{-1}, respectively. In the seawater/mud mixture (sample B) the AOR during the first feeding cycle was only 4.9 mg N L^{-1} day^{-1}. It increased three-fold to 15.1 mg N L^{-1} day^{-1} during the second feeding cycle after day 42 (Fig. 4.1b).

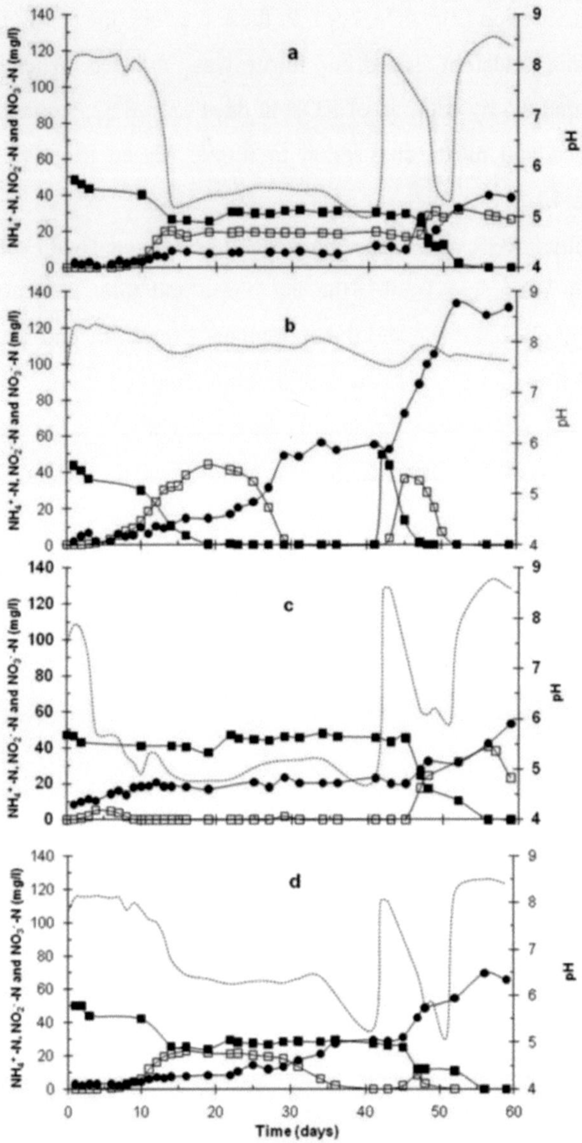

Figure 4.1 Initiation of nitrification of ammonia (50 mg N L^{-1}) by autotrophic nitrifiers from different samples of seawater/mud. Sample sources were a) Hafen Büsum (seawater), b) Town Norden (seawater), c) Town St. Peter-Ording (brine water) and d) Town St. Peter-Ording (seawater).Enrichment reactors with 2 L h^{-1} aeration rate. At days 42 and 52, the pH was raised with KOH. In assay b), ammonia was added for a second time at day 42. The symbols represent: closed squares (■), ammonia; open squares (□), nitrite; closed circles (●), nitrate and dotted line (---), pH. The incubation temperature was 20–23°C

4.2 Polyethylene/clay sinter lamellas and porous ceramic rings as substrata for biofilm formation in fixed-bed reactors (FBRs) for an aerobic treatment of saline wastewater

Seawater sample A from Hafen Büsum was used as an inoculum of FBR A (polyethylene/clay sinter lamellas) and FBR B (porous ceramic rings). Sample A was preferred over sample B because it did not contain any mud and its AOR of 11.5 mg N day^{-1} was only slightly lower than that of sample B. Both FBRs were filled with 2.1 L seawater sample A (Table 4.1). After addition of 104 mg N L^{-1} ammonia, FBR A and FBR B were operated under batch conditions for 5 weeks to acclimatize the nitrifying microorganisms. Figure 4.2 shows the course of ammonia and nitrite concentrations in both reactors during acclimatization.

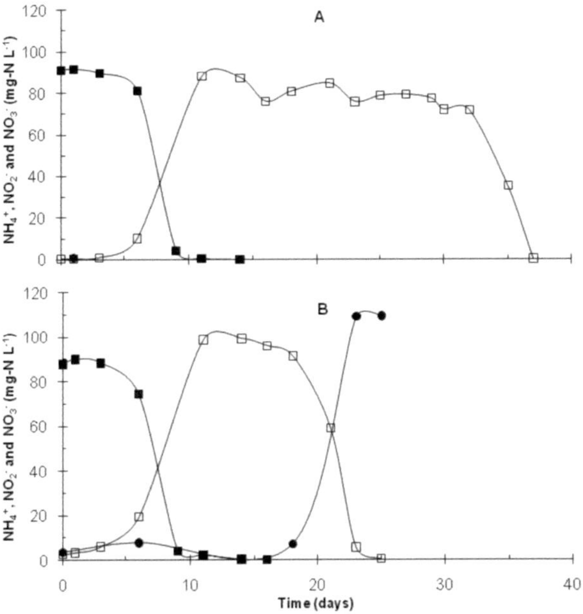

Figure 4.2 Course of ammonia (closed squares,■), nitrite (open squares,□) and nitrate (closed circles,●) in FBR A (Polyethylene/clay sinter lamellas) and B (porous ceramic rings) during acclimatization

The initially measured ammonia concentration was about 90 mg N L^{-1}, which was lower than the added ammonia concentration (104 mg N L^{-1}). The initial disappearance of ammonia might have been caused by ammonia absorption in the supporting media. Five days lag phase for biomass in both reactors were needed before ammonia oxidation to nitrite started. The ammonia oxidation rates in both reactors, once oxidation began, were almost the same. There was a double as long time requirement in FBR A (20 d lag phase) before nitrite was further oxidized to nitrate than in FBR B (10 days).

After acclimatization, a synthetic seawater medium was supplemented with NH_4Cl as ammonia source and both FBRs were running for more than 300 days. Four phases (see Table 3.1) were distinguished, concerning the HRT, ammonia loading rate (ALR) and installation of an external water recirculation to improve the mixing intensity for establishment of better conversion rates.

The courses of ammonia, nitrite, nitrate and pH in FBR A and FBR B with time are shown in Figures 4.3 and 4.4, respectively.

During the initial phase a (days 0–55, Figure 4.3) in FBR A ammonia was completely oxidized and more nitrite was formed than nitrate. Due to malfunction of the pH titrator at day 10, causing an increment of the pH to 9.1, 60 mg N L^{-1} ammonia remained in the reactor effluent and nitrite and nitrate concentrations decreased. Restoration of the full AOR was obtained after the exchange of the pH probe within the next 10 days. From day 45 onwards all ammonia was oxidized to the expected stoichiometric amount of nitrate without intermediary nitrite accumulation (Figure 4.3a).

In phase b (days 56–155, Figure 4.3b), the HRT was reduced from 1.25 to 1 day, thereby the ALR increased from 83 to 104 mg N L^{-1} day^{-1}. Under steady-state conditions, complete ammonia oxidation continued. Due to a failure of the pH probe at days 66 and 88, the pH fell below 6.5. Ammonia accumulation started immediately, but after replacement of the pH probe degradation of ammonia resumed. The diaphragm of pH electrodes was often clogged by carbonate or other salt precipitates, which lead to the failure of the pH measurement. Cleaning of the pH electrodes by immersion in 0.1 M HCl solution for one half hour was then conducted to overcome the failure.

From days 112 and 140, the aeration rate in FBR A increased drastically due to a leak of the internal aeration system. Higher air flow rates caused higher shear forces, and this led to biofilm detachment from the carrier material. Small flocks on the conical bottom of reactor and in the bottle for effluent collection indicated visible detachment of the nitrifying biofilm. The disturbance of the fixed biomass resulted in an incomplete ammonia oxidation and some nitrite accumulation (Figure 4.3b, day 112).

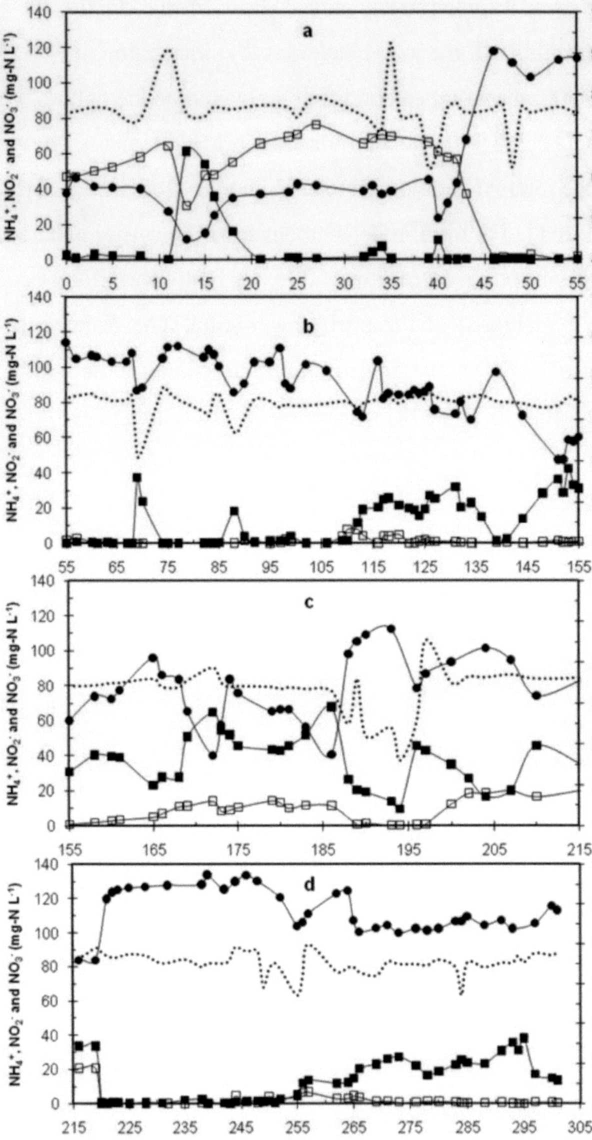

Figure 4.3 Ammonia, nitrite, and nitrate concentrations during continuous operation of FBR A (support material: polyethylene/clay sinter lamellas) for conditions of Table 3.1. Days 0–55, phase a; days 55–155, phase b; days 155–215, phase c; days 215–305, phase d. Symbols: closed squares (■), ammonia; open squares (□), nitrite; closed circles (●), nitrate; dotted line (---), pH. The incubation temperature was 20–23°C at an air flow rate of 2.5 L h^{-1}

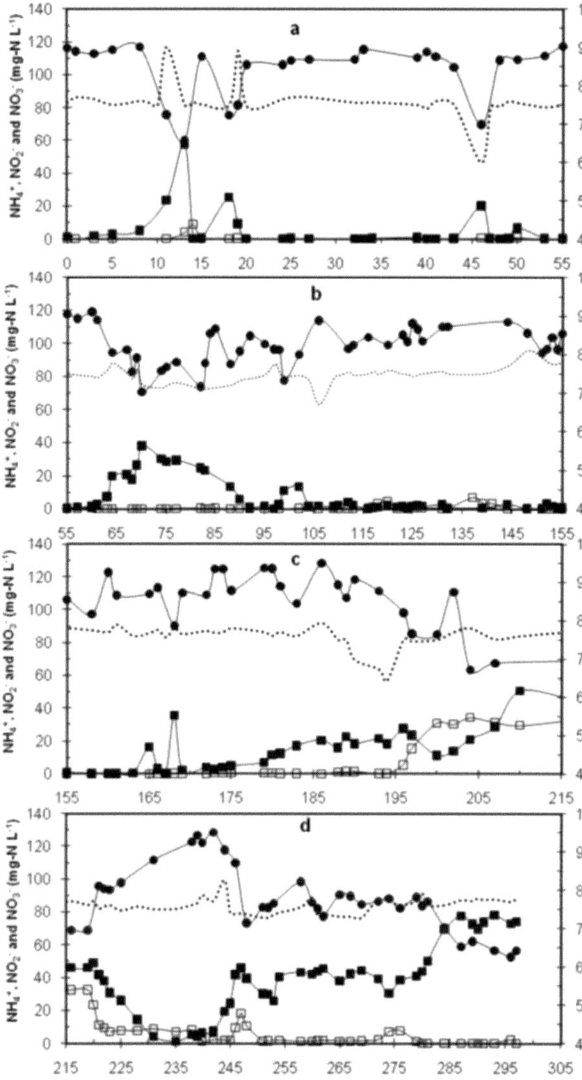

Figure 4.4 Ammonia, nitrite, and nitrate concentrations during continuous operation of FBR B (support material: porous ceramic rings) for conditions of Table 3.1. Days 0–55, phase a; days 55–155, phase b; days 155–215, phase c; days 215–305, phase d. Symbols: closed squares (■), ammonia; open squares (□), nitrite; closed circles (●), nitrate; dotted line (---), pH. The incubation temperature was 20–23°C at an air flow rate of 2.5 L h^{-1}

In phase c (days 156 – 215, Fig 4.3c), the ALR was further increased to 130 mg N L^{-1} day^{-1} by increasing the ammonia concentration in the medium. Ammonia was no longer oxidized completely and some nitrite accumulated. Due to problems with the stability of pH probes in the saline environment, the titration system was switched off and the medium was buffered by addition of $NaHCO_3$ from day 187 to 194. Ammonia oxidation to nitrate improved significantly, but precipitates in the sodium bicarbonate buffered medium led to clogging of the inlet tubes and thus to a disturbance of high rate reactor operation. After re-installation of the pH titration system and the omission of sodium bicarbonate from day 195 onwards, ammonia oxidation improved, leading to around 20 mg N L^{-1} nitrite and up to 100 mg N L^{-1} nitrate, respectively (Figure 4.3c).

In phase d of FBR A operation, an effluent recirculation of 8 L h^{-1} (1.25 m h^{-1} upstream velocity) was installed to improve the mixing of the medium. Immediately after starting recirculation of the reactor content, ammonia was completely oxidized without intermediate formation of nitrite (days 219–250, Figure 4.3d). From day 250 to the end of reactor operation, 20 – 40 mg N L^{-1} ammonia remained in the effluent, but no nitrite accumulated. The detached biomass visibly settled in the conical bottom of the reactor after applying the recirculation.

Figure 4.4 shows the results of FBR B during continuous operation. Contrary to FBR A, the biofilm on the porous ceramic rings of FBR B could oxidize all ammonia to nitrate in phase a (Figure 4.4a) after starting the continuous operation. Short-term instabilities in saline medium due to a failure of the pH probe occurred in FBR B and caused a pH decrease during days 8, 15 and 46, which led to an incomplete oxidation of ammonia until repair. However during the disturbance, no significant amounts of nitrite accumulated.

An increase of the ALR to 104 mg N L^{-1} day^{-1} in phase b of FBR B led to an increase of the ammonia and a decrease of the nitrate concentration from days 60 to 70 in the reactor effluent but no nitrite accumulated. All ammonia was oxidized to nitrate however from day 90 onwards (Figure 4.4b). A further increase of the ALR to 130 mg N L^{-1} day^{-1} in phase c did not affect the nitrifying performance of the reactor initially (Figure 4.4c). However, after day 180, ammonia was no longer oxidized completely, nitrite accumulated and nitrate decreased. Replacing the titration unit by a bicarbonate buffered medium (days 187–194) could not stop the decrease of nitrate formation (Figure 4.4c; days 186–194).

Aside of that, clogging of the inlet tubes due to precipitated sodium carbonate occurred also in FBR B. When the bicarbonate was omitted and the pH titrator re-installed, ammonia oxidation and nitrate formation for a short time improved but accumulating nitrite led to a reversal of the improvement (Figure 4.4c). To overcome this decreasing efficiency, an external recirculation was installed in phase d after day 216 to improve mixing. This led to a short recovery of ammonia and nitrite oxidation (Figure 4.4d, days 219–240), but similar as in FBR A some biofilm detachment occurred and was visible by the sludge, that accumulated at the bottom of the reactor. From days 240 to 280, 40 mg N L^{-1} ammonia remained in the reactor. Later on to the end of reactor operation the ammonia concentration was 80 mg N L^{-1}, however no nitrite accumulated.

AORs and NORs of FBR A and FBR B on the respective days in Table 4.2 during almost one year are summarized in Table 4.2. For determining the AORs and NORs in FBR A and FBR B, the ammonia supply was stopped 1 day before rate analyses to obtain a complete depletion of ammonia and nitrite in the reactors. Either NH_4Cl or KNO_2 was then added to each reactor to reach a final concentration of 60 or 30 mg N L^{-1}, respectively. AORs and NORs were determined from the slope of the consumption curves, and a surface area-related nitritation and nitration rate for FBR A and FBR B was calculated (Table 4.2).

Table 4.2 Surface area-specific ammonia and nitrite oxidation rates in FBR A and FBR B

Days	AOR (mg N $m^{-2}day^{-1}$)		Days	NOR (mg N $m^{-2}day^{-1}$)	
	FBR A	FBR B		FBR A	FBR B
48	nd	nd	48	274	230
64	306	156	65	230	226
84	176	51	85	386	236
113	145	156	114	n.d.	n.d.
173	312	199	174	n.d.	n.d.
215	207	197	216	363	182
217	502	197	218	547	181
257	295	112	258	484	210
284	272	149	285	n.d.	n.d.
295	218	82	296	525	214

n.d. = not determined.

NORs were always higher than AORs, except for day 64 in FBR A, and for day 215 and day 217 in FBR B, where nitrite accumulated due to an unknown disturbance (Figure 4.4c and d). The surface area-related AORs and NORs in FBR B with porous ceramic rings were always lower than those in FBR A with polyethylene/clay sinter lamellas, except for day 113 for AORs. The NOR, especially in FBR B, was relatively stable, whereas the AOR showed a higher fluctuation with time. After starting the external recirculation (phase d; Table 3.1), the AOR in FBR A was increasing more than two-fold and the NOR more than 1.5-fold, whereas the AOR and NOR in FBR B were not improving (Table 4.2; day 217). There was no positive short-term effect in FBR B as in FBR A, where the AOR at first increased and then slowly decreased to the initial rates before recirculation, but the capacity for nitrite oxidation remained high. The NOR was not increased by a better homogenization through recirculation in FBR B (Table 4.2).

4.3 Effect of salinity fluctuations in the fixed-bed reactors

4.3.1 Effect of higher and lower salinities in FBRs C, D and E with porous ceramic rings as fixed-bed material for a starting salinity of 3.5 % sodium chloride

The previous experiments showed that stable nitrification in saline wastewater can proceed in FBRs with porous ceramic rings or polyethylene/clay sinter lamellas as support materials. Complete ammonia oxidation without nitrite accumulation occurred at almost all reactor operational conditions. Only technical failures, such as malfunction of the pH electrode or a highly increased turbulence during intensive aeration caused incomplete oxidation.

Industrial wastewater with a high salinity often fluctuates in volume and also characteristically in its salinity. Salinity fluctuations however, might significantly influence the performance of nitrification in wastewater reactors.

In this experiment, fixed bed reactors with porous ceramic rings were fed with synthetic saline wastewater with a changing salinity (see Table 3.2 and Figure 3.3). The saline wastewater, which was used for this experiment, contained 60 mg NH_4^+-N per L as in previous studies. The results of the experiment are presented in Figure 4.5.

In phase I (days 0–15, Figure 4.5) in all reactors a complete ammonia oxidation to the expected stoichiometric amount of nitrate without nitrite accumulation was obtained. The alkalinity added to the influent medium with a ratio of alkalinity consumption to ammonia oxidation of 7.1 mg $CaCO_3$/mgNH_4^+ was sufficient to maintain a constant pH of 8±0.2. The reactors were run at a steady state for about two weeks.

In phase II (days 16–27), the salinities of FBRs C, D and E were changed successively from 3.5% to 0.5, 1 and 2% NaCl, respectively by supplying a medium with the mentioned salinity during continuous operation. The salt concentration in the reactors decreased gradually and – due to the short HRT -

after about 3 days (approximately 6 volume replacements) the salt concentration was close to that of the influent.

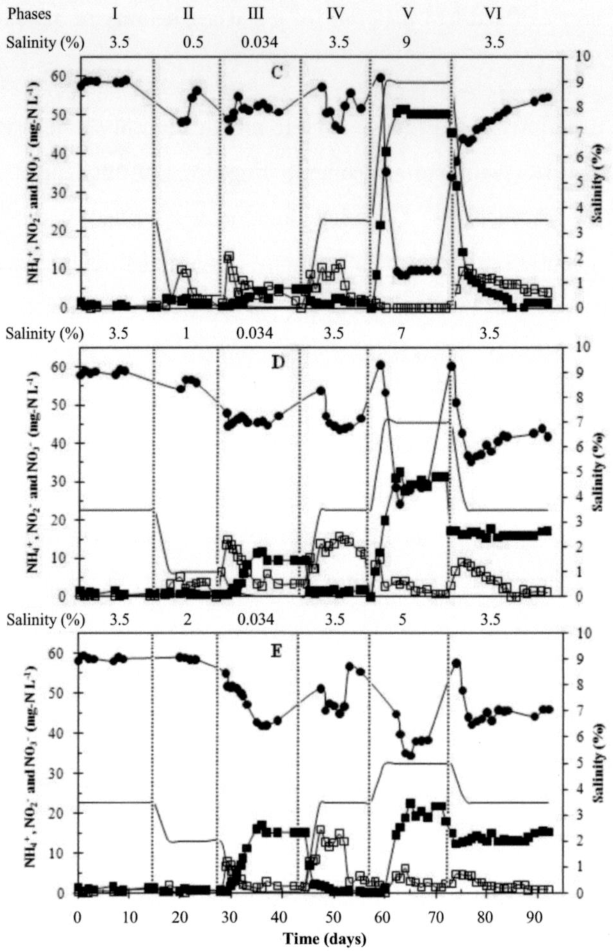

Figure 4.5 Concentration of ammonia, nitrite, and nitrate during continuous operation of FBRs C, D and E at conditions of Table 3.2. Symbols: closed squares (■), ammonia; open squares (□), nitrite; closed circles (●), nitrate; solid line (–), alkalinity. The incubation temperature was 20–23°C

In this phase, ammonia in FBR C with 0.5 % NaCl was still almost completely oxidized while a complete ammonia oxidation proceeded in FBRs D and E with the higher salt concentrations. Nitrite accumulation was detected in FBR C (10 mg N L^{-1}) and in FBR D (5 mg N L^{-1}). After 7 days, complete oxidation of ammonia and nitrite was achieved in all reactors. This results show that the biofilm in the FBRs had an ability to completely oxidize ammonia and nitrite even at salt concentrations of only 0.5 % NaCl after several days of adaption.

Therefore in phase III (days 28-43), the salt concentration of the influent medium was further reduced to 0.034%, the salinity of tap water in the laboratory. This was done to check

whether halophilic or halotolerant nitrifying biomass can survive at tap water minerals concentration and to analyze how the extent of salinity changes influences the nitrifying activity. Salinity was decreased from previously 2, 1 and 0.5 % to 0.034 % in FBRs C, D and E, respectively.

Salt concentration of only 0.034 % immediately led to incomplete ammonia oxidation. In all reactors, ammonia in the effluent increased gradually to 5 mg N L^{-1} in FBR C, 10 mg N L^{-1} in FBR D and to 15 mg N L^{-1} in FBR E and then remained relatively constant after about 5 days. The ammonia oxidation efficiency in FBR E, where the highest decrement of salinity was applied (from 2 % to 0.034 %), was lower than that in the other FBRs.

Nitrite concentrations increased immediately when salinities were reduced, but within about 5 days decreased again to almost zero mg N L^{-1}. The highest nitrite accumulation in the reactors was about 15 mg N L^{-1} in FBRs C and D and 5 mg N L^{-1} in FBR E. The low salinity of 0.034 % clearly reduced metabolic activities of ammonia oxidizing bacteria. The bacteria either could not cope with the osmotic shock and then decayed at the low salt concentration or the bacteria had survived, but with a limited activity.

In phase IV (days 44-57) all reactors were fed with the initial salt concentration (3.5 %) to test the recovery ability of the biomass. It could be shown that the ammonia removal efficiency of 100 % was achieved almost immediately in the FBRs; however 10-15 mg N L^{-1} of nitrite accumulated during the beginning of this phase.

It is expected that ammonia oxidizing and nitrite oxidizing bacteria respond differently to a decrease or an increase of salinity. Therefore, in a first attempt in phase V (days 58-74), the salinity of the wastewater was increased from 3.5 % to 9 % (FBR C), 7 % (FBR D) and 5 % (FBR E).

It could be shown that salinity increases led to a notable decrease of the ammonia oxidation in FBR C by a factor 6.5 (from 100 % to 16 %), in FBR D by a factor of 2 (from 100 % to 55 %) and in FBR E by a factor of 1.5 (from 100 % to 70 %). Nitrite concentrations of 3-4 mg N L^{-1} were detected in the effluent of FBRs D and E.

The ability for metabolic recovery of the biomass after a feeding period with the higher salinity for two weeks was analyzed in phase VI (days 75-85), in which the salinity in the medium was brought back into the initial concentration of 3.5 % NaCl. The efficiency of ammonia removal increased in all reactors, which could be seen by a lower ammonia concentration in the effluent.

However a complete ammonia oxidation occurred only in FBR C, while the ammonia concentration in the effluent of FBR D was 15 mg N L^{-1} and the ammonia concentration in FBR E was 12 mg N $L^{-1}$and remained constant until the end of experiment.

The decrease of salinity from 9, 7 or 5 % to 3.5 % NaCl resulted in a nitrite accumulation for a short while immediately after changing the salinities in phase VI. Similar observations were also made during phases II and III, where the salinity was also decreased.

- *Ammonia and nitrite oxidation rates in FBRs*

Performances of nitrifying bacteria in fixed bed reactors were clearly affected by a decrease or an increase of the salinity of the applied wastewater. Immediately after salinity changes it could be elucidated, how nitrifying bacteria respond to fluctuations of salinity. In the experiment there were however several situations in which it was difficult to come to a clear and scientifically correct conclusion. Nitrite accumulation in phase V, for instance, could have been caused by a changing salinity and/or by ammonia accumulation, whereas in phase II, when no ammonia was detected in the effluent of all reactors, the extent of salinity changes might have had no effect on ammonia oxidizing bacteria. The other possibility would be that the changes had an effect on conversion rates, but not strong enough, so that the ammonia oxidizing bacteria could still consume all ammonia at the decreased salinity. Therefore, the ammonia concentration in the effluent remained zero as in the previous phase.

Ammonia and nitrite oxidation rates were measured to round up the results of experiments during continuous operation of the reactors.

AORs and NORs were experimentally determined at the end of each phase by shifting the reactors for a short while into batch wise operation. The results of the rate measurements are presented in Figure 4.6 and Figure 4.7.

Figure 4.6 shows that ammonia and nitrite concentrations decreased relatively uniform from the initial concentration of 20 mg N L^{-1} ammonia (for AOR) and 15 mg N L^{-1} nitrite (for NOR) to about 2 mg N L^{-1}. This indicated that AOR and NOR were constant in this concentration range.

The measurements of the AORs in the three reactors resulted in almost the same rates for NaCl from 0 – 3.5%, while NORs were different (Figure 4.7). The NORs were always higher than the AORs in all reactors. Moreover, in the first measurement the NOR was more than twice as fast as the AOR. The decrease of salinity from 3.5% to 2, 1 and 0.5%, respectively, did not influence the AORs. A decrease of the AORs was found after the salt concentrations were brought down to 0.034%. A significant decrease of the NORs in the three FBRs occurred already at the salinity decrease from 3% to 1%, but the NORs were still higher than the AORs.

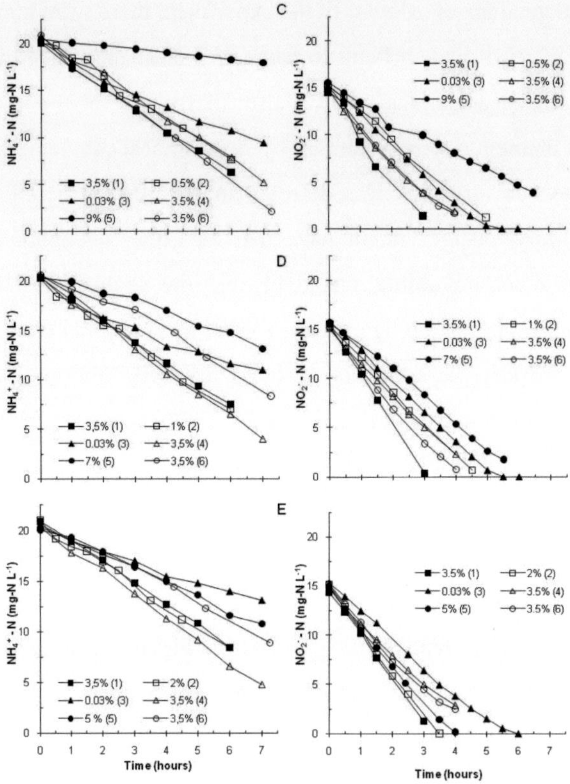

(a) (b)

Figure 4.6 Concentration of ammonia (a) and nitrite (b) in batch wise operated FBRs C, D and E for starting NaCl concentrations of 0.034 % and final NaCl concentrations of either 5, 7 or 9%

The decrease of salinity from 3.5 % via 2 % and 1 % to 0.034 % resulted in a decrease of NORs to 71 % and 75 % of the initial rates, respectively, while the final NOR was only 55 % of the initial rate, when the salinity was decreased from 3.5 % via 0.5 % to 0.034 %. Similar rates were obtained, when NORs were measured at 0.5 % and 0.034 %.

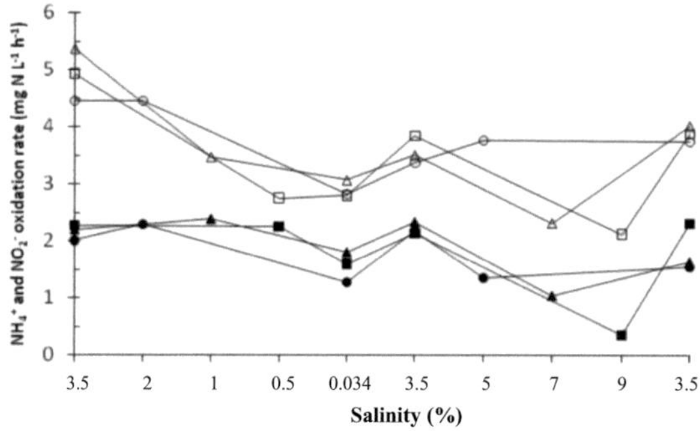

Figure 4.7 Ammonia oxidation rates (AOR) and nitrite oxidation rates (NOR) in FBRs C, D and E. Symbols: closed squares (■), AOR in FBR C; open squares (□), NOR in FBR C; closed triangles (▲), AOR in FBR D; open triangles (△), NOR in FBR D; closed circles (●), AOR in FBR E; open circles (○), NOR in FBR E

The recovery ability of a halophilic or halotolerant nitrifying consortium that was cultured in a medium without salt for some time and then again exposed to a medium with 3.5 % NaCl for ammonia oxidation by AOB was better than that for nitrite oxidation by NOB. When the reactors were supplied again with saline medium (3.5 % NaCl), the AORs increased up to the initial rates, while the NORs increased only to 60 – 80 % of the initial rates.

An increase of the salinity from 3.5 % to 5, 7 or 9 %, respectively resulted in decreasing AORs and NORs with increasing NaCl concentrations, with the exception of NOR in FBR E, which increased slightly. The highest decrease was found in FBR C, in which the salt concentration was increased from 3.5 % to 9 %. The recovery ability of NOB at decreasing salinity was better than that of AOB. When the influent salt concentration in all reactors with very high salt concentrations was reduced to 3.5 % in all reactors, NORs increased up to the original rate which was measured for 3.5 % NaCl before increasing the salinity to 5, 7 or 9 % NaCl.

4.3.2 Nitrifying activity in polyethylene/clay fixed bed reactors FBRs F, G, H and I for salinity changes from 3.5 % to lower and higher % values of NaCl

In the FBR with porous ceramic rings as carrier material salinity fluctuations affected the activity of the nitrifying biomass. During the salt fluctuations the biomass was able to maintain AORs and NORs above 50 % of the initial rates, except for the highest salinity of 9 %. In order to relate the results to the supporting material, effects of salinity fluctuations on nitrifying biomass was examined in detail (see Figure 3.4) with polyethylene/clay as support material. Salt concentrations during operational phases of the FBRs were adjusted as stated in Table 3.3 and the course of NH_4^+, NO_2^-, NO_3^-, pH and salinity is presented in Figure 4.8.

In phases I, II and III an almost complete ammonia oxidation without nitrite accumulation was obtained in FBRs F, G and H. The positive trend continued in phase IV (days 36-49) when the salinity of three reactors was 0.034 %. Only in FBR H nitrite accumulation was detected. However, after 5 days a gradual increasing ammonia concentration in all reactors and in the effluent was found. Ammonia concentrations at the end of this phase were 20, 40 and 40 mg N L^{-1} for FBRs F, G and H, respectively.

In phase VI (days 64-79) the salinity of the medium was changed from 3.5 % to 9 % (FBR F), 7 % (FBR G) and 5 % NaCl (FBR H), respectively. Ammonia concentrations increased steadily to 30 mg N L^{-1} (FBR F) and to 15 mg N L^{-1} (FBR G) and remained constant until the end of this phase.

In phase V (days 50-63) and VII (days 80-90) the salinity in medium was changed back to the salinity of 3.5 % at the start of the experiment after the reactors were fed with lower or higher salinity for about one month or half a month, respectively. In both phases, ammonia concentrations in the reactors decreased immediately to zero, representing an achievement of 100 % ammonia removal. In other words, ammonia oxidizing bacteria have the ability for a good recovery from salt shocks.

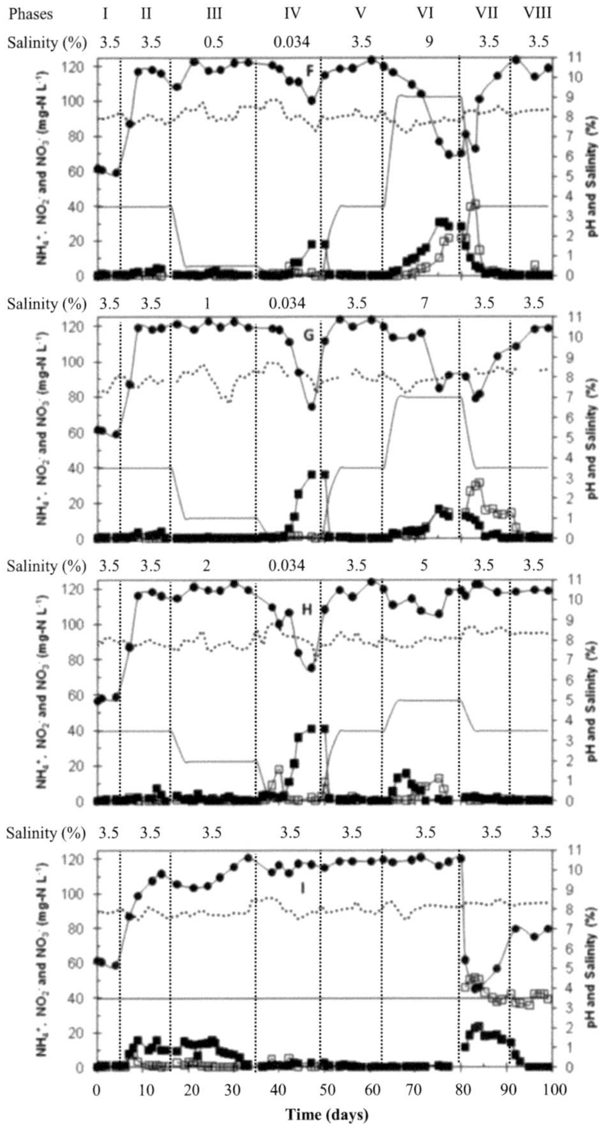

Figure 4.8 Ammonia, nitrite, and nitrate concentrations during continuous operation of FBRs F, G, H and I (all with polyethylene/sinter lamellas) for the conditions mentioned in Table 3.3. Symbols: closed squares (■), ammonia; open squares (□), nitrite; closed circles (●), nitrate; horizontal dotted line (---), pH; solid line (–), salinity. The incubation temperature was 20–23°C

A different result was obtained concerning nitrite oxidizers. The nitrite concentration in the end of phase III and V was zero mg N L^{-1}. The nitrite concentration at the end of phase VI was lower than that in the beginning of phase VII (FBRs F and G).

The nitrite concentration in FBRs F and G increased in the beginning of phase VII from 20 mg N L^{-1} to 40 in FBR F and from 15 mg N L-1 to 30 mg N L^{-1} in FBR G. The concentrations, then, decreased to zero in 5 days (FBR F) and 10 days (FBR G).

FBR I was a control reactor which was fed with synthetic seawater containing 3.5 % salinity during the whole experiments. When the ammonia concentration in the influent was doubled from 60 mg N L^{-1} (phase I) to 120 mg N L^{-1}, about 15 mg N L^{-1} of ammonia was detected in the effluent for about 20 days. The increased ammonia concentration also led to a temporary nitrite accumulation in the beginning of the phase II.

A steady-state with 100 % of ammonia and nitrite removal efficiency could be maintained in the control reactor for 45 days (35 to 80 days). Although no operational parameters had been actively changed on day 80, in the next phase about 20 mg N L^{-1} of ammonia and 40 mg N L^{-1} of nitrite accumulated, along with a decrease of nitrate concentrations. The ammonia decreased gradually to zero in about 15 days, whereas the nitrite was not completely oxidized to nitrate until the end of the experiment.

- *Ammonia and nitrite oxidation rates in FBRs F, G, H and I*

Figure 4.9 shows AORs and NORs in FBR F, G and H and Figure 4.10 shows AORs and NORs in FBR I (control reactor).

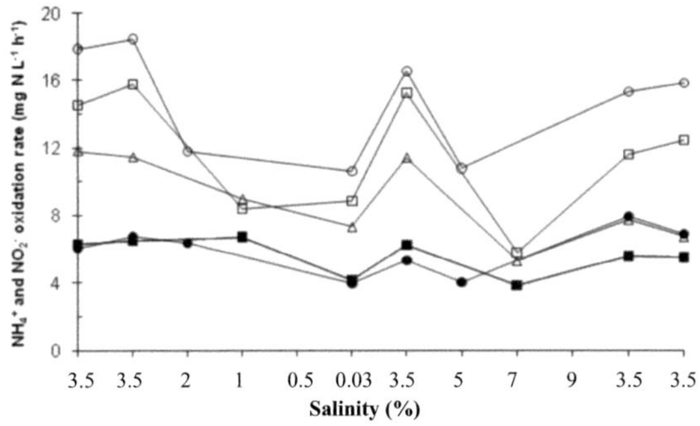

Figure 4.9 Ammonia oxidizing rates (AORs) and nitrite oxidizing rate (NORs) in FBRs F, G and H. Symbols: closed squares (■), AORs in FBR F; open squares (□), NORs in FBR F; closed triangles (▲), AORs in FBR G; open triangles (△), NORs in FBR G; closed circles (●), AORs in FBR H and open circles (○), NORs in FBR H

Similar as the ammonia oxidation rates in FBRs C, D and E (Figure 4.7), the AORs in FBRs F, G and H were much more uniform for changing salinities as compared to the NORs, which decreased with higher or lower salinities than 3.5 % NaCl. Furthermore, the NORs were always higher than the AORs.

Doubling of the ammonia concentration in the medium from 60 mg N L^{-1} (phase I) to 120 mg N L^{-1} (phase II) did not result in a significant change of the AORs and NORs. Only the NORs in FBR F and H slightly increased. The AORs remained constant, although the salinity in the medium was decreased from 3.5 % to 2, 1 or 0.5 %. The AORs, finally, decreased after the salinity was further decreased to 0.034 %. The highest decrease of the AOR was found in FBR F when the salinity was decreased from 3.5 % to 2 % and then to 0.034 %. In this reactor the lowest increase of the AOR was obtained when the medium was brought back to 3.5 % salinity.

On the other hand, a decrease of the NORs occurred already when the salinity was decreasing from 3.5 % to 2 %. The NOR at a salinity of 0.5 % was the same as that at a salinity of 0.034 %, a result that was obtained in FBRs C, D and E (Figure 4.7).

It can also be seen that the increase of the salinities from 3.5 to 5, 7 or 9 % resulted in a proportional reduction of the AORs and NORs, but NORs remained higher than AORs for all salt concentrations that were applied.

AORs and NORs in FBR I (control reactor) increased slightly with time (Figure 4.10). However, the measurement of rates at phase VI (days 78 and 79) revealed a significant decrease of the NOR. Afterwards, NORs were lower than AORs until the end of the experiment. This fact was also confirmed with a nitrite accumulation in the effluent of FBR I during continuous operation (Figure 4.8, phase VII).

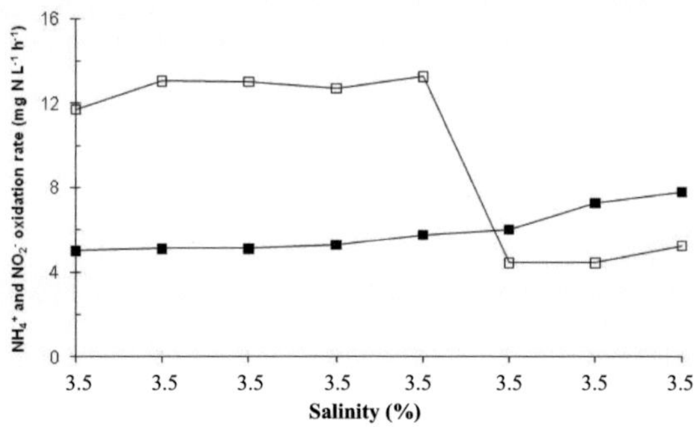

Figure 4.10 Ammonia oxidizing rates (AORs) and nitrite oxidizing rates (NORs) in FBR I (control reactor). Symbols: closed squares (■), AOR in FBR I; open squares (□), NOR in FBR I

4.3.3 Nitrification of ammonia in polyethylene/clay fixed bed reactors (FBRs J, K, L, M and N) under non-halophilic conditions (0.035 % salinity start concentration) during increases of the salinity up to 10.5 % NaCl

The previous experiments (section 4.31 and 4.32) showed that nitrification in the fixed bed reactors that were inoculated with biomass originating from sea water in Hafen Büsum functioned relatively good in a salinity range from 0.034 % to 7 %. In this experiment, similar fixed bed reactors with polyethylene/clay sinter lamellas as supporting material were started with non saline surface water taken from the region of Büsum. The water was "activated" by nitrate addition for initiation of nitrification before it was used as an inoculum.

Ammonia, nitrite, nitrate and pH were monitored during operation of the reactors with media that contained successively the concentrations of NaCl of Table 3.4 and results of the experiments are presented in Figure 4.11.

Reactors J, K, L, M and N were acclimatized during a batch mode (phase I, days 0–8) by addition of 60 mg NH_4^+-N L^{-1}. Ammonia oxidation began immediately and no ammonia was left after 5 days. During ammonia oxidation in phase I nitrite accumulated to more than half of the concentration, that was theoretically maximally possible at a 1:1 stoichiometry. After depletion of ammonia, the nitrite was further oxidized to nitrate. The oxidation rates of nitrite accelerated faster than those of ammonia in the beginning of this phase.

From phase II (days 9-22) onwards, all reactors were continuously operated. Complete ammonia and nitrite oxidation was achieved in the beginning of this phase, when the ammonia concentration in the medium was 60 mg N L^{-1}. The positive performance in all reactors continued when the ammonia concentration was doubled to 120 mg N L^{-1}. A higher ammonia concentration was not applied in this experiment, because it would need more alkalinity for keeping a ratio of 7.1 mg $CaCO_3$: 1 mg NH_4^+-N. Alkalinity in high concentration causes the formation of precipitated salts, which could cover the biofilm on the supporting material and decease its activity.

In phase III of the experiment (days 23-36) the salinity in FBRs J, K, L and M was increased from 0.034 % to 0.5, 1, 2 and 3.5 %, respectively, while the salinity of 0.034 % in FBR N (control reactor) was kept constant until the end of the experiment.

Complete ammonia oxidation could still be maintained in all reactors. However, nitrite accumulation was detected in all reactors except for FBR J. The highest nitrite accumulation was detected in FBR M, where the salinity was increased from 0.034 % to 3.5 %.

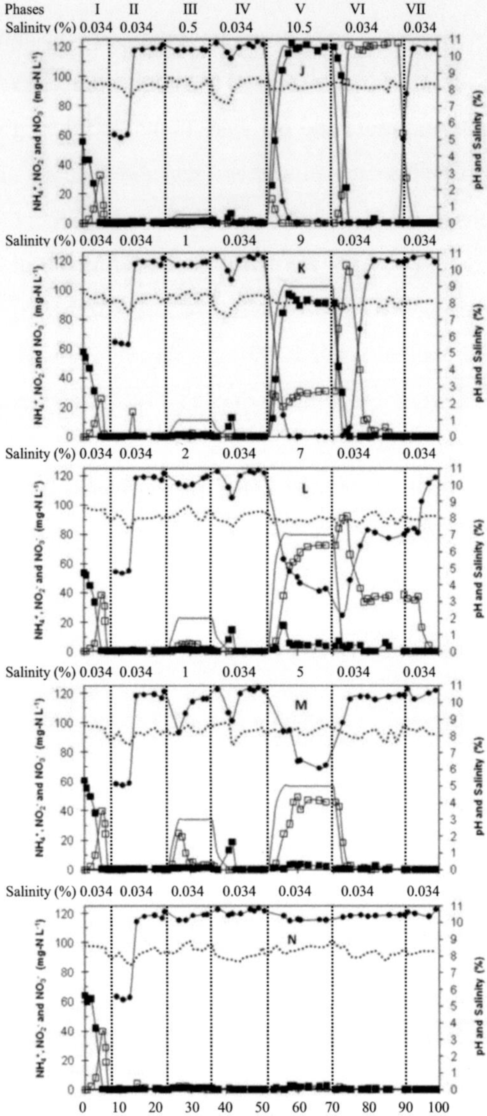

Figure 4.11 Ammonia, nitrite and nitrate concentrations during continuous operation of FBRs J, K, L, M and N (all with polyethylene/clay sinter lamellas, non-halophilic inoculums) for conditions mentioned in Table 3.4. Symbols: closed squares (■), ammonia; open squares (□), nitrite; closed circles (●), nitrate; horizontal dotted line (---), pH; solid line (–), salinity. The incubation temperature was 20–23°C

A much lower nitrite accumulation was obtained in FBRs K and L, in which the salinity had been increased from initially 0.034 % to 1 or 2 % only. Thus, the nitrate-forming NOB were more affected by the increasing salinity than the nitrite-forming AOB.

In contrary, ammonia and not nitrite, was detected in the effluent, when the salinity was brought back to 0.034 % (phase IV, days 37-51) to examine a recovery ability of the biomass. Ammonia was detected only for 2 – 4 days at the beginning of this phase, which indicated an immediate recovery of the biomass.

To check for the tolerance of the nitrifying bacteria of real high salinities, in phase V (days 52-69) the salinity was increased from 0.034 % to 10.5 % (FBR J), 9 % (FBR K), 7 % (FBR L) and 5 % (FBR R). The concentrations 9, 7 and 5 % NaCl in FBRs K, L and M were chosen to enable a direct comparison with previous experiments (section 4.3.1 and 4.3.2), while 10.5 % (FBR J) salinity was applied as an additional higher salinity.

In FBR J with 10.5 % salinity in the medium ammonia in the reactor was not oxidized and it left the reactor in the same concentration as in the influent, indicating a complete inhibition of ammonia oxidizing bacteria. Whether nitrite oxidizing bacteria were active could not be seen, since no nitrite was formed by the AOB. At steady state conditions in FBR K ammonia, nitrite and nitrate concentrations were 90, 30 and 0 mg N L^{-1}. It can be said that AOB could still oxidize 30 mg N L^{-1} ammonia of the 120 mg N L^{-1} that were supplied, whereas NOB were not active. In this reactor partial nitrification as required for the Anammox-process might be maintained.

In FBRs L and M, the nitrifying biomass still had the ability to oxidize all supplied ammonia. Nitrite accumulation (70 mg N L^{-1} in FBR L and 40 mg N L^{-1} in FBR M), however, was detected in the medium until the end of phase V.

In phase VI (days 70-89) the initial salt concentration of 0.034 % was re-established in all reactors. In FBR J, where the salinity had to be decreased from

10.5 % to 0.034 % only nitrite was found in the reactor, indicating that AOB recovered fast but NOB apparently could not be reactivated. Contrary results were obtained in the previous phase, in which AOB had no activity after an increase of the salinity from 0.034 % to 10.5 %.

The relatively quick recovery of AOB and/or NOB was also obtained in FBRs K and M. Unexpectedly in FBR L, where the decrease of salinity was not as drastic (from 7 % to 0.034 %) compared to that in FBR K (from 9 % to 0.034 %), nitrite accumulation was still detected. This means that the recovery of NOB in FBR L was slower than in FBRs K and M, which might be related to nitrite accumulation in the previous phase.

In the prolongation phase VII (days 90-99) it turned out that nitrite accumulation in FBRs J and L lasted on for 20-25 days.

In FBR N (control reactor) a nitrification efficiency of 100 % was always maintained during continuous operation of the reactor.

- *Ammonia and nitrite oxidation rates in FBRs J, K, L, M and N*

AOR and NOR measurements of FBRs J, K, L, M and N were conducted at the end of each phase and results are shown in Figure 4.12 and 4.13.

AORs and NORs were more variable with increasing salinity in this experiment where a non-halophilic or halo-tolerant nitrifying population was used as inoculums in contrary to the previous experiments with halophilic or halotolerant populations at the start of the experiment. For salt concentrations not exceeding 1 % the NORs were higher than the AORs in all reactors (Figure 4.12). Compared to AORs during batch wise mode (phase I, 0.034%), the rates during continuous operation (phase II, 0.034%) were slightly faster, while NORs during both conditions were relatively the similar and not different.

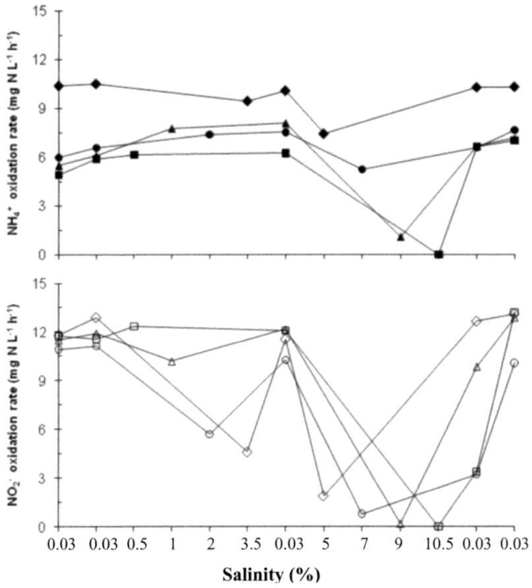

Figure 4.12 Ammonia oxidation rates (AORs) and nitrite oxidation rates (NORs) in FBRs J, K, L and M (all with polyethylene/clay sinter lamellas) with a non-halophilic or non halotolerant inoculum. Symbols: closed squares (■), AORs in FBR J; open squares (□), NORs in FBR J; closed triangles (▲), AOR in FBR K; open triangles (△), NORs in FBR K; closed circles (●), AORs in FBR L; open circles (○), NORs in FBR L; closed diamonds (◆), AORs in FBR M; open diamonds (◇), NORs in FBR M

A salinity increase from 0.034 % to 0.5, 1, 2 and 3.5 % did not influence the AORs. A salinity effect on AOR was however found after the salt concentration was increased beyond 3.5%. A salinity effect on NOR was already found at a salinity of only 1%. NORs were more drastically inhibited at 2% and 3.5% NaCl concentration in the medium and the rates were only about half of the initial rates.

After the medium in the reactors was brought back to the initial salt concentration (0.034% - phase IV), the AORs and NORs were similar as the rates in the phase II, indicating that the effect of an increasing salinity was temporary.

The increase of salinities from 0.034% to 5, 7, 9 and 10.5 % NaCl resulted in significant inhibitions of the AORs and NORs. At salt concentrations of 10.5 %, neither a nitrite nor an ammonia oxidation activity was seen (FBR M). At a salt concentration of 9% no nitrite oxidation activity was seen and the ammonia oxidation activity was still close to zero (FBR L). For an increase of the salinity from 0.034% to 5 (FBR J) or 7% (FBR K) AORs stayed at about 70% of their initial rates.

It looked like, that NORs were more sensitive than AORs for higher salinities, but the inhibiting effect surprisingly seemed to be temporary.

In the control reactor FBR N AORs and NORs continuously increased with operation time (100 days) by a factor of 1.35 for AOR and 1.25 for NOR (Figure 4.13).

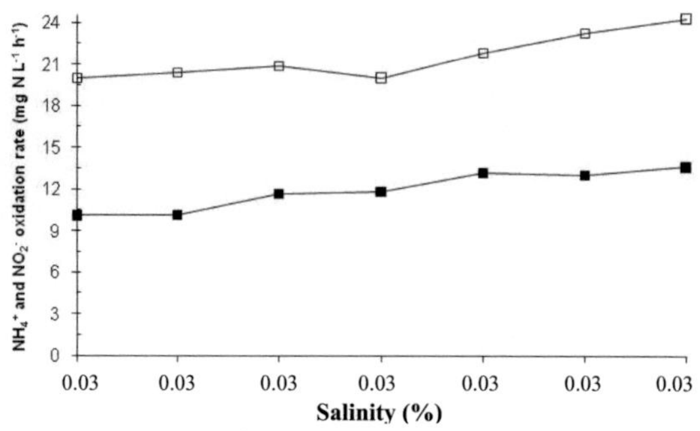

Figure 4.13 Ammonia oxidation rates (AORs) and nitrite oxidation rates (NOR) in FBR N (control reactor) during 100 days of operation. Symbols: closed squares (■), AOR in FBR N; open squares (□), NOR in FBR N

4.4 Effect of a sudden or gradual increase of the ammonia concentration in polyethylene/clay fixed bed reactors (FBRs O, P and Q)

In FBR A and FBR B, a better homogenization of synthetic seawater was achieved with an external recirculation. The synthetic seawater from the top of the reactor was pumped through silicon tubes (length 1.5 m, diameter 1 cm) into the reactor bottom (see section 3.3.1.1).It was found, after about 2 month of external recirculation, that a thin biofilm had been formed on the inside-wall of the tubes.

Nitrifying activity of this biofilm was observed in batch assays by scratching off the biofilm and suspending it into 250 ml-glasses. The glasses were filled with synthetic seawater (3.5% of salinity, 100 mg N L^{-1} ammonia or 20 mg N L^{-1} nitrite) and were mixed and aerated. Results of ammonia and nitrite measurements from the batch assays can be seen in Figure 4.14.

Ammonia was immediately oxidized to nitrite after the first and the second feeding. After almost 100 hours the nitrite concentration was 100 mg N L^{-1}, exactly the stoichiometric amount that could be generated from ammonia oxidation, which indicated no activity of nitrite oxidizer. This was confirmed in the other batch assay, which was fed only with nitrite (Figure 4.14b) and where no nitrite oxidation was seen.

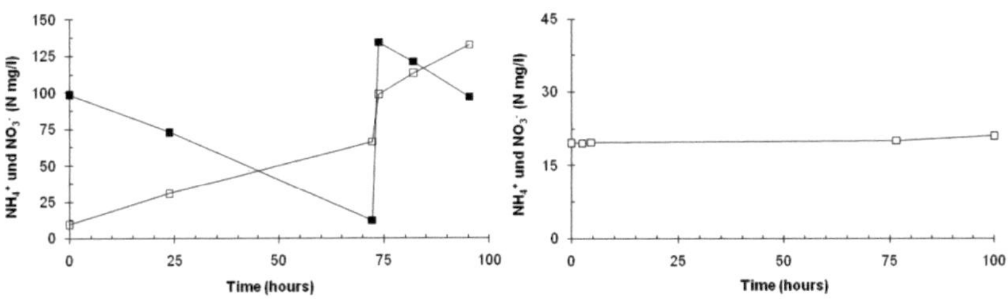

(a) (b)

Figure 4.14 Concentration of ammonia and nitrite during the test of ammonia- and nitrite oxidizing activity of the biomass that was taken from the inner surface of the tubing for external recirculation of FBRs A and B media. Symbols: closed squares (■), ammonia; open squares (□), nitrite. Second feeding after 75 hours

The biofilm from the tubes was also applied as an inoculum in continuously run fixed bed reactors with polyethylene/clay as supporting material (see section3.3.1.5). For 20 days the reactors were operated in batch mode to acclimatize the biomass and after that continuous operation with increasing ammonia concentrations was started (see Table 3.5). The result of ammonia, nitrite and pH monitoring can be seen in Figure 4.15.

After the first (100 mg NH_4^+-N L^{-1}) and the second (150 mg NH_4^+-N L^{-1}) feeding ammonia was immediately oxidized in all reactors. The total ammonia fed into FBR P was only about 170 mg NH_4^+-N L^{-1} due to a leakage in the reactors during the second feeding.

Nitrite produced by ammonia oxidizers has not been oxidized until day 10. This confirms the result of the batch assays, where no activity of nitrite oxidizer was obtained by the biofilm biomass that was removed from the inner surface of silicon tubes.

Until the end of the experiment, for FBR Q (control reactor) and until the end of phase II for FBRs Q and P, the biomass in the reactors could maintain a complete ammonia and nitrite removal. Incomplete ammonia oxidation was found at day 40, 60, and 68 and seemed to be due to a pH decrease. The instability of the pH occurred along with visible precipitation of salts in small diameter inlet pipes of the reactors that make up the alkalinity.

To solve the pH problem, Tris(hydroxymethyl)aminomethane buffer (TRIS buffer) was applied. However based on the result of a test measurement, TRIS interfered with spectrophotometrical measurement of ammonia (data not shown). Thus, instead of TRIS-buffer application, bigger inlet pipes were used and the pH in the reactors was manually adjusted with KOH or H_3PO_4.

In FBR O, the complete ammonia and nitrite oxidation continued during phase III (days 73-89), when the ammonia concentration was increased by a factor of 2 (from 40 mg N L^{-1} to 85 mg N L^{-1}). However a further increasing of the ammonia concentration from 85 mg-N L^{-1} to 125 mg N L^{-1} in phase IV (days 90-105) failed to maintain the positive trend. Ammonia as well as nitrite was detected in the effluent.

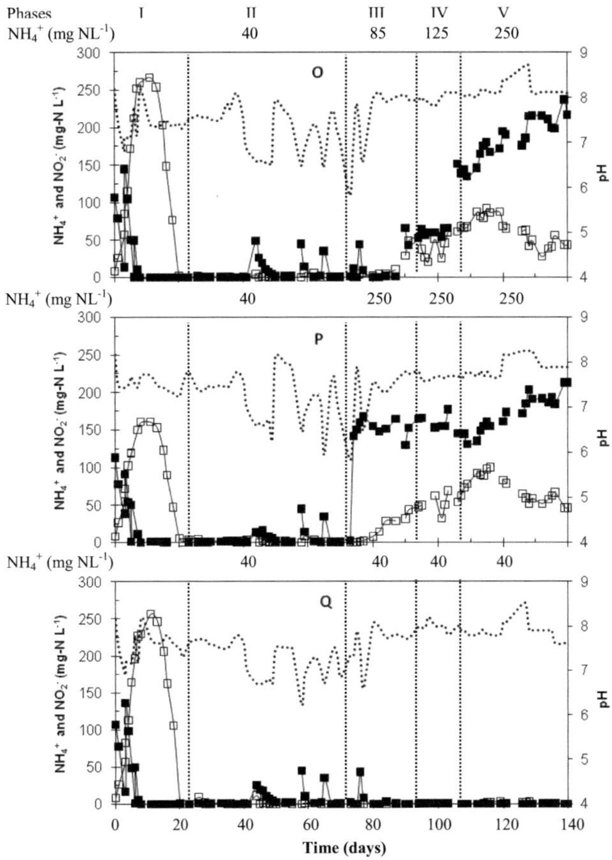

Figure 4.15 Ammonia, nitrite, and pH during continuous operation of FBRs O, P and Q (porous ceramic rings) for conditions of Table 3.5. Symbols: closed squares (■), ammonia; open squares (□), nitrite; horizontal dotted line (---), pH

The biomass in FBR O could oxidize on average about half (62.5 mg N L^{-1}) of the total ammonia supply in the influent (125 mg N L^{-1}) to about 62.5 mg N L^{-1} nitrite. The total amount of nitrogen from ammonia and nitrite in the effluent was the same as the nitrogen content of ammonia in the influent, indicating that no nitrate was produced. These would be optimal condition for the Canon process, the combination of partial nitrification and Anammox in a single aerated reactor.

After ammonia concentration was increased from 125 mg N L^{-1} to 250 mg N L^{-1} in phase V (days 106-140), the biomass could oxidize 100 mg N L^{-1} of ammonia for 10 days, but then the oxidation ability decreased gradually until the end of the experiment. Also, no nitrite oxidation occurred during the phase, which was indicated by no nitrate in the effluent (data not shown).

In FBR P, ammonia was increased by a factor of about 6 from 40 to 250 mg N L^{-1} in phase III and was kept at this concentration until the end of the experiment. On average 150 mg N L^{-1} of ammonia was detected in the effluent or, in the other words, biomass in the reactor could oxidize 100 mg N L^{-1} ammonia for 40 days. This ability was the same with FBR O, when salinity was increased from 125 to 250 mg N L^{-1} in phase V. Nitrite accumulation increased steadily, corresponding to a decreasing nitrate concentration. Finally, after 40 days no nitrate was formed any more.

- *Ammonia and nitrite oxidation rates in FBRs O, P and Q*

Nitrification rate measurements for FBRs J, K, L and M as well as for FBR N were conducted at the end of each phase and the results are shown in Figure 4.12 and 4.13.

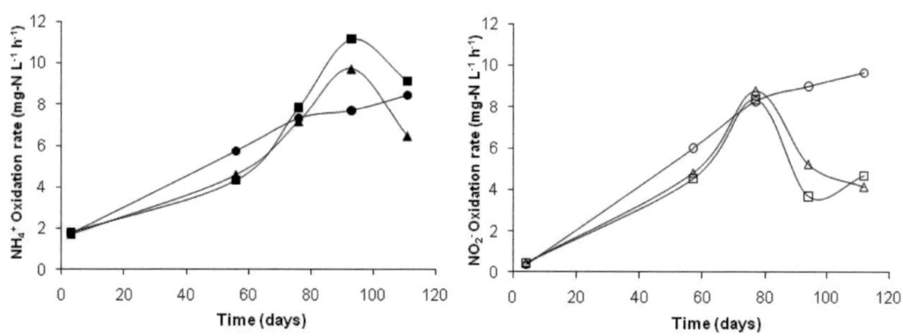

(a) (b)

Figure 4.16 Ammonia oxidizing rates (a) and nitrite oxidizing rates (b) in the FBRs O, P and Q. Symbols: closed squares (■), AOR in FBR O; open squares (□), NOR in FBR O; closed triangles (▲), AOR in FBR P; open triangles (△), NOR in FBR P; closed circles (●), AOR in FBR Q; open circles (○), NOR in FBR Q

In FBR Q (control reactor) only in phase I the AOR was higher than the NOR. Then for 53 days (days 3 to 56) AORs increased by a factor of 3, while NORs increased by a factor of 20. From day 54 to the last rate measurement (55 days) NORs were always higher than AORs.

In FBR O, AORs increased slightly faster than those of the control reactor until phase IV, when 125 mg N L^{-1} of ammonia was present in the influent. The AORs however decreased after the ammonia concentration was increased to 250 mg N L^{-1}.

In the contrary, an increase of ammonia from 40 to 250 mg N L^{-1} in FBR P in phase III resulted in a higher AOR. A second measurement of the AOR at an ammonia concentration of 250 mg N L^{-1} again resulted in a higher rate.

The NORs increased in FBRs O and P until phase III, in which the ammonia concentration in the influent of FBRs D and P was 85 and 250 mg N L^{-1}, respectively. A further increase of the ammonia concentration in FBR O or prolonged incubation under such concentrations in FBR P resulted in a drastic decrease of NORs.

4.5 Biomass growth in porous ceramic rings

Four distinct stages are believed as developmental processes of biofilm formation: reversible attachment, irreversible attachment, maturation and detachment. Measurement of oxidation rates of biomass that is growing attached on carrier material, which has been incubated for increasing time intervals in FBRs until steady state nitrification is obtained, might represent the biofilm developing process.

In such an experiment, fresh porous ceramic rings were put inside FBR B (section 3.3.1.1) that has been running continuously for 540 days. After 1, 4, 12, 20, 33 and 50 days, some of the added “fresh” porous ceramic rings were taken out and the removal of ammonia and nitrite by the biomass that was growing attached to the porous ceramic rings hitherto was measured separately in small-scale batch assays (see Figure 3.8).

Ammonia, nitrite, and nitrate concentrations as well as pH in FBR A during the experiment are shown in Figure 4.17a; while the oxidation rates of attached biomass on porous ceramic rings after different incubation time are presented in Figure 4.17b.

By operating FBR B at a HRT of 2 days, a DO of above 5 mg L^{-1}, a pH of 5 – 8 and an ammonia concentration of 50 mg L^{-1} in the influent steady state conditions with no ammonia and nitrite in the effluent were achieved during the experiment. It was expected that the steady state conditions allow a direct comparison of biomass activity.

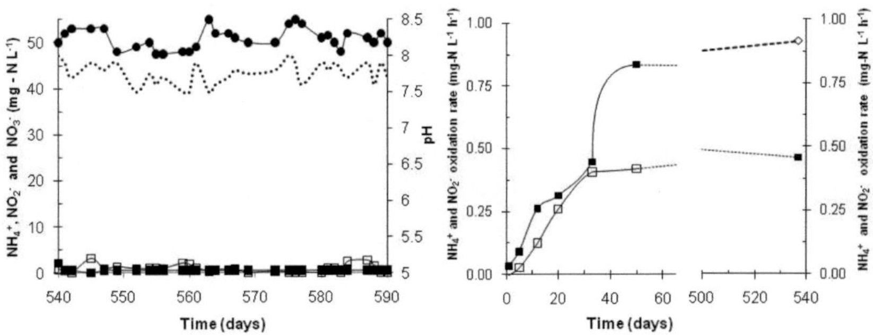

(a) (b)

Figure 4.17 (a) Ammonia, nitrite, and nitrate concentrations during continuous operation of FBR B, days 540 – 590. (b) Ammonia and nitrite oxidizing rates by developing biofilm in support materials after the indicated time of incubation. Symbols: (a) closed squares (■), ammonia; open squares (□), nitrite; closed circles (●), nitrate; dotted line, pH. (b) Closed squares (■), AOR; open squares (O), NOR

It can be seen that after 50 days of incubation AORs were always higher than NORs. Additionally, ammonia oxidation proceeded faster and was finished after 1 day incubation, indicating that ammonia oxidizers attach or grow faster in the porous ceramic material than nitrite oxidizers. The AORs increased unevenly until day 50, whereas NORs increased constantly until day 30 and then remained at steady values.

In the beginning of the experiment, the oxidation rates of the biofilm that was attached on "old" porous ceramic rings were also determined. The "old" porous ceramic rings were already in the respective FBR for 540 days. Results with the mature biofilm on these carriers show that NOR was twice as high after 540 than NOR after 50 days of incubation of fresh carriers, while the AOR measured on the carriers after 537 days was somewhat lower than the AOR after 50 days. It should be noted, that during 540 days of incubation steady state conditions could not always be maintained.

4.6 Distribution of ammonia and nitrite oxidation capabilities in the biofilm on ceramic rings or polyethylene/clay sinter lamellas taken from the same height of a FBR

After several months, growth of a biofilm on polyethylene/clay (FBR A) and porous ceramic rings (FBR B) was visible (Figure 4.18). The biofilm seems to grow not in a uniform manner. An improper mixing in biological processes normally results in a non-homogeneous medium distribution in all parts of a reactor. Although substrates and nutrients may not be distributed homogeneously in all parts of the reactors the nitrifying biofilm looked more like a sediment on horizontal parts of the porous glass rings.

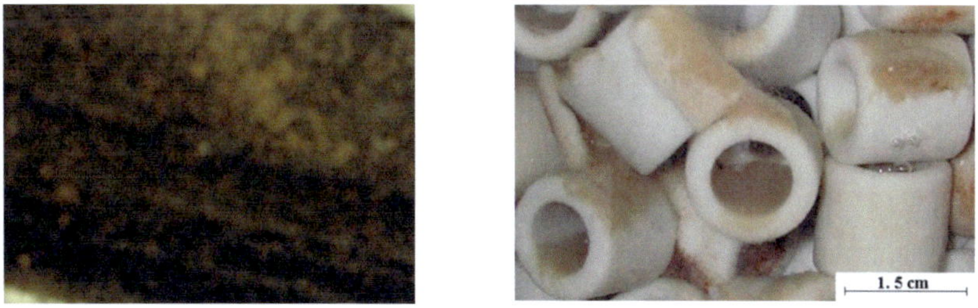

(a) (b)

Figure 4.18 Biofilm attached to supporting material (a) Polyethylene/clay from FBR A (b) porous ceramic rings from FBR B

In attached growth reactors many factors may be affecting the distribution of biomass on single supporting porous glass rings or on different supporting glass rings.

Ammonia and nitrite oxidation rates by the biofilm on several porous ceramic rings that have been taken from FBR B at the same height were measured separately in small reactors.

The results (Figure 4.19) show that the activity of the biofilm population was different by factor of 1.5 and 2 for ammonia and nitrite oxidation, respectively.

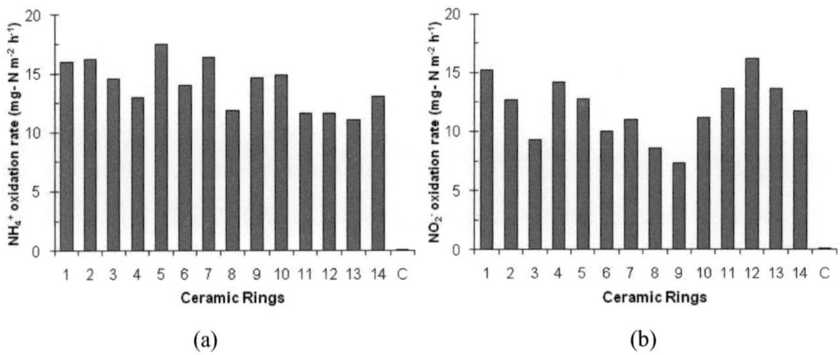

Figure 4.19 Ammonia (a) and nitrite (b) oxidation rates by biofilm bacteria on porous ceramic rings

4.7 The effect of azide and allylthiourea (ATU) on ammonia or nitrite oxidation rates by the biofilm in porous ceramic rings

Ammonia and nitrite oxidizing bacteria are responsible for nitrification. However, it is difficult to determine separately the activities of both types of bacteria because they exist in colonies closely associated. They respond however differently to inhibitors of nitrification. The use of allythiourea, known as a specific inhibitor of nitritation and sodium azide, known as a specific inhibitor of nitratation makes a separate measurement of the two reactions of nitrification possible.

The oxidation rate measurement was conducted using biofilm nitrifiers on porous ceramic rings in small reactors. After AORs and NORs were determined without addition of allythiourea or natrium azide in first measurements, the specific inhibitors were added into the small reactors with concentrations ranging from 0.02 – 2 mg L^{-1} for allythiourea and 0.1 – 80 μg L^{-1} for natrium azide. AORs and NORs without and with inhibitors were measured (Figure 4.20).

It can be seen that allythiourea was a specific and potential inhibitor of nitritation. Addition of only 0.2 mgL^{-1} of allythiourea led to a complete inhibition of the activities of AOB (Figure 4.20a), whereas addition of allylthiourea up to 2 mgL^{-1} had no effect on the nitrite oxidation activity.

In contrast, sodium azide seemed to be no specific inhibitor for nitritation. Addition of 40 μg L^{-1} of the substance, indeed, resulted in complete inhibition of the activity of NOB. However it also brought about a decrease of the nitratation activity.

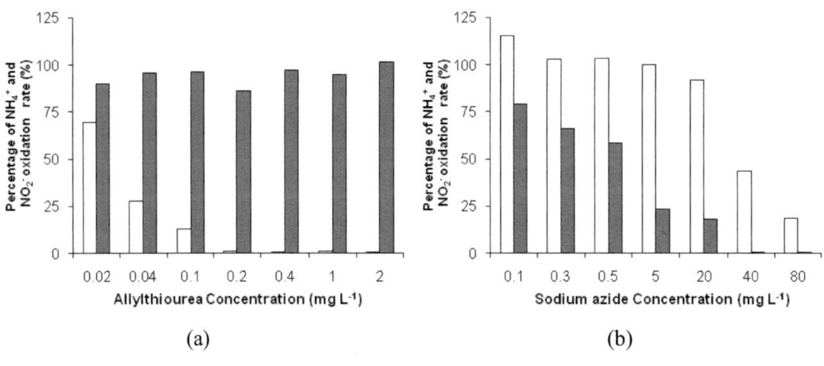

Figure 4.20 Ammonia oxidation rates in percent at different concentrations of allythiourea (a) and natrium azide (b). Symbols; open bar, AOR; closed bar, NOR

4.8 The effect of salinity on ammonia or nitrite oxidation rates by biofilm nitrifiers

In continuously operating FBRs (section 4.3.1-3) effects of a salinity change on ammonia and nitrite oxidation rates by biofilm nitrifiers were already observed. The measurement of the ammonia and nitrite oxidation rates however, were conducted after the biofilm nitrifiers were already incubated at a new salt concentration for about 2 weeks. The bacteria might have had enough time to adapt to the new environment.

In this experiment, biofilm nitrifiers in porous ceramic rings were used as inocula and put into the small reactors. AORs and NORs were determined at three times. The first and third measurements were conducted at 3.5 % salinity in the medium and the second measurement was performed at a different salinity (see Table 3.7). The result of the experiment can be seen in Figure 4.21.

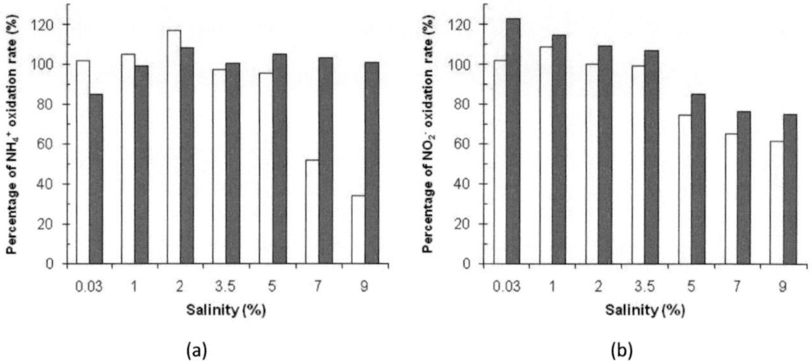

Figure 4.21 Ammonia (a) and nitrite (b) oxidation rates in percent at different salinities. Symbols; first measurement = 100%, open bar, second measurement; closed bar, third measurement

It can be seen that the rates from the second and third measurements of AOR or NOR in the control reactor at 3.5% salinity gave the same rates (about 100%) as the first measurement. It indicated that the measurement method was duplicable and no significant amount of biofilm was detached during medium drainage.

It was obtained that AORs were not affected by a salinity change from 3.5 % to 5, 2, 1, or 0.03 % and NORs from 3.5 to 2, 1, and 0.03 %. The AORs and NORs by biofilm nitrifiers, adapted to 3.5 % salinity, seemed to be more sensitive towards salinity increases than towards a salinity decrease.

Only about 60% and 30 % activity of AOB and NOB, respectively, was found when a salinity of 9 % was applied in assays. However AOB and NOB have a good recovery after reducing the salinity back to 3.5 %, where the recovery of AOB was slightly higher than that of NOB.

4.9 The effect of salinity on the ammonia and nitrite oxidation rates by suspended nitrifiers

A biofilm in a fixed bed reactor could function as a protector for microorganisms to a shock change of their environment, such as a shock load or shock pH. Therefore, a reactor with attached biomass on support materials might be more stable than a reactor with suspended biomass, especially if it is operated at fluctuating environmental conditions.

In the last experiment, biofilm nitrifiers were used for the measurement of the effect of shock salinity changes on AOR and NOR. It seemed that AORs and NORs were only influenced by increases of the salinity and not or much less by decreases of the salinity.

Instead of the biofilm nitrifiers, in this experiment nitrifiers taken from a suspension reactor (CSTR and SBR, see section 3.3.2.8) that was run with 3.5 % salinity were used as inocula.

- *Suspended nitrifying bacteria from a CSTR*

After centrifugation, the pelleted biomass was put into 50 ml medium, containing 60 mg NH_4^+-N L^{-1} or 25 mg NO_2^- -N L^{-1} with different salinities (0.03; 0.5; 1; 2; 3.5; 5; 7; 9 %). The recovery of the nitrification activity at 3.5 %

salinity after incubation with the above mentioned higher or lower salt concentrations was also examined.

The decrease of ammonia concentrations accompanied by nitrite accumulation during the first rate measurements (CSTR) are presented in Figure 4.22a and b. The data of Figure 4.22 were used to calculate the ammonia oxidation rates during the experiment, and the results are presented in Figure 4.23.

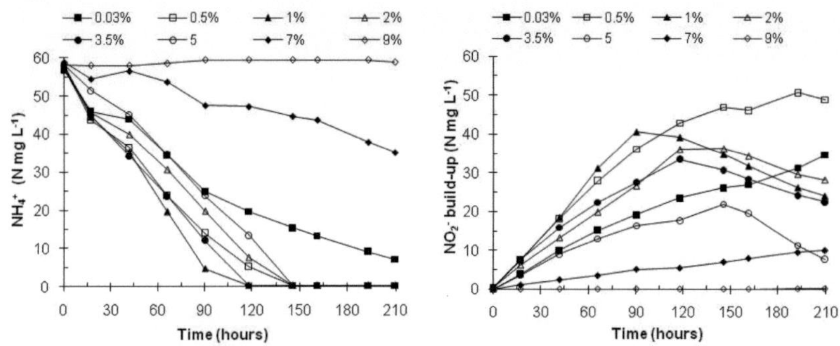

(a) (b)

Figure 4.22 Ammonia concentrations (a) and nitrite build-up (b) during the measurement of AORs by suspended bacteria that were pre-incubated at 3.5% salinity and transferred to media with higher or lower NaCl concentration

The highest rate surprisingly was obtained at a salinity of 1 % NaCl. Increasing salinity decreased the activity of AOB more than decreasing the salinity. In addition, the rate of ammonia decrease in the reactor with 3.5 % NaCl (control reactor) was comparable with that at 0.5 % NaCl. The most severe inhibition effect was obtained at salinities of 7% and 9%.

By comparing oxidation rates of the first and second measurements (Figure 4.23a) it can be seen that the effect of the salinity change to very low or high salinities (0.03, 5, 7 and 9% NaCl) on AORs were reversible.

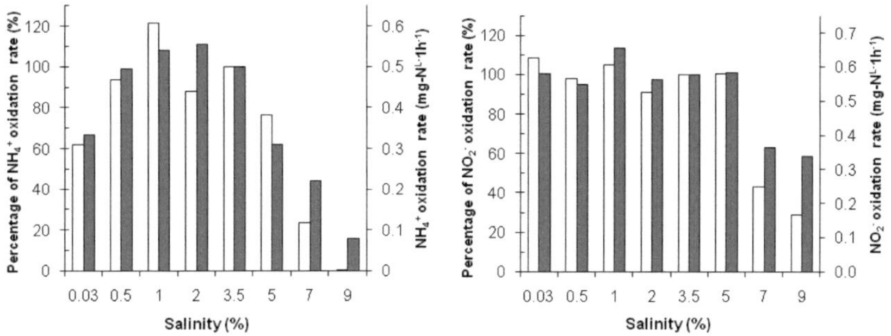

(a) (b)

Figure 4.23 Ammonia (a) and nitrite (b) oxidation rates at different salinities by suspended nitrifiers. Symbols; open bar, first measurement at changed salinities; closed bar, second measurement after reestablishment of the original salinity (3.5% NaCl)

For NORs (Figure 4.23b), the activity of NOB in the biofilm could cover a salinity range from 0.03 % to 5 %. In such a range of salinities, NORs were relatively constant at values obtained in the control reactor with 3.5 % salinity. An irreversible inhibition effect was found when the salinity was increased from 3.5 % to 7 and 9 %.

Separate measurements of nitritation and nitratation in control assays with 3.5% salinity shows that the NORs by suspended nitrifying bacteria were slightly higher than the AORs. However some nitrite was detected during ammonia oxidation measurement.

Accumulation of nitrite during nitrification (Figure 4.22b) indicated that the bacterial activity that generated nitrite (nitritation) was higher than the activity of the nitrite-consuming NOB (nitratation). It can be seen that nitrite accumulation occurred in all assays during incubation, including the control assays with the originally 3.5 % salinity. No nitrite was detected in the assay with 9 % of salinity, since the AOB were apparently completely inhibited at this high salt concentration. The ammonia oxidation rates, nitrite accumulation rates and the ratio of both rates as well as the maximum detected nitrite concentration are summarized in Table 4.3.

As nitritation and nitratation proceed simultaneously, the ratio of NAR/AOR (NAR = nitrite accumulation rate) presents how fast nitrite is produced by AOB and how fast it will be oxidized to nitrate by NOB. The NAR/AOR ratio of 1 represents a process in which only nitritation exists, the ratio of 0 indicates than the nitratation rate is at least as high as or higher than the nitritation rate. A high NAR/AOR ratio at decreasing salinity indicates a maximum nitrite accumulation, for example in the assay with 0.5 % of salinity, a condition that should be avoided.

Table 4.3 Ammonia conversion rates and maximum nitrite accumulation during nitrification by suspended bacteria that were pre-grown at 3.5% NaCl concentration

Salinity %	AOR $mgNL^{-1}h^{-1}$	AOR:AOR control	NAR* $mg NL^{-1}h^{-1}$	NAR/AOR	Max of Nitrite Accum. $mgNL^{-1}$
0.03	0.31	61	0.23	0.75	34
0.5	0.47	93	0.42	0.89	50
1	0.61	120	0.46	0.76	40
2	0.44	86	0.3	0.69	36
3.5	0.50	100	0.35	0.69	33
5	0.38	78	0.20	0.52	21
7	0.12	22	0.05	0.45	9
9	0	0	-	-	0

* = Nitrite Accumulation Rate

In case of the assay with 1 % of salinity, the AOR was higher than in the control assay (3.5%). However, this positive result leads to a higher NAR:AOR ratio than the ratio obtained from the control assay.

Contrary to that a low NAR:AOR value and a lower maximum nitrite accumulation was obtained in the assay with 5 % NaCl. The AOR, however, was lower than in the control.

It can be also be said that the low salinity resulted in a higher NAR:AOR ratio and vice versa.

- *Suspended bacteria from a SBR*

Previous experiments proofed that nitrifiers in a biofilm and suspended nitrifiers that originated from a marine ecosystem could maintain a relatively high activity during a shock by decreasing the salinity. Vice versa, the ability of non-halophilic nitrifiers to "function" at increasing salinities is also an interesting subject.

Results of determination of the effect of salinity on the AORs and nitrite accumulation by non-halophilic suspended nitrifiers are presented in Figure 4.24.

The same AORs were obtained from assays with 0.03 % or the control assay, and at 1 % of salinity, indicating no effect of salt concentration up to 1 % on AORs. A severe effect of the salinity on AORs was obtained in assays over a salt concentration of 3 % and over 7 % salinity a complete inhibition of the AORs was detected.

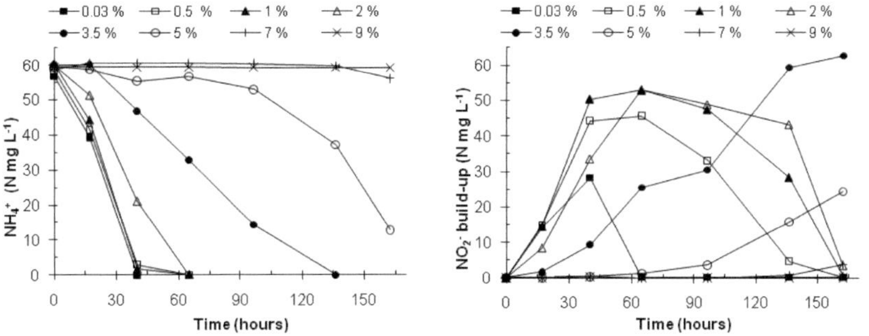

(a) (b)

Figure 4.24 Ammonia concentration (a) and nitrite accumulation (b) during the measurement of ammonia oxidation rates by suspended bacteria (0.03%)

In Figure 4.25, the result of rate measurements of ammonia and nitrite oxidation in percent is presented. It seem that NOBs were more sensitive to salinity than AOBs, where AORs in assays with 0.5 and 1 % were about half of the rate of control assays. The recovery test for both measurements was not done.

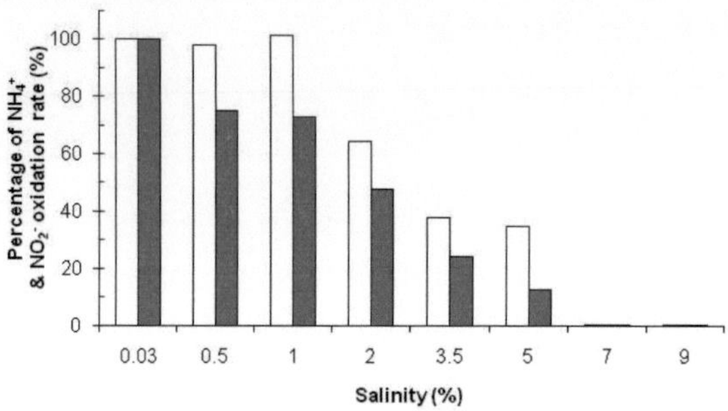

Figure 4.25 Ammonia (open bar) and nitrite (closed bar) oxidation rates at different salinity by suspended bacteria (0.03%)

As in experiments with halotolerant nitrifiers, nitrite accumulation was detected during this experiment, as summarized in Table 4.4. It can be seen that the NAR:AOR ratio and the maximal nitrite accumulation increased when the salinity was increased until 3.5 %. As said before, increasing salinity from 0.03 % to 0.5 and 1 % had no effect on AOR. However this increase resulted in a higher NOR:AOR and a much higher maximum of nitrite accumulation compared to the control assay (0.03 %).

Table 4.4 Rate of nitrite accumulation during nitrification by non-halophilic suspended bacteria

Salinity %	AOR $mgNL^{-1}h^{-1}$	AOR:AOR control	NAR* $mgNL^{-1}h^{-1}$	NAR:AOR	Max of Nitrite Accum. $mgNL^{-1}$
0.03	1.37	100	0.82	0.60	28
0.5	1.34	98	1.08	0.80	45
1	1.39	101	1.21	0.87	52
2	0.88	64	0.79	0.89	52
3.5	0.52	38	0.40	0.76	62
5	0.48	35	0.21	0.45	25
7	0	0	-	-	0
9	0	0	-	-	0

* = Nitrite Accumulation Rate

4.10 The effect of free ammonia (NH_3) on ammonia and nitrite oxidation rates by biofilm-forming nitrifiers

Besides its function as a substrate, high concentrations of uncharged ammonia might inhibit nitrification. Biofilm-carrying porous ceramic rings from FBR B were put into small cylindrical glass reactors as an inoculum (see Figure 3.8).

Oxidation rate measurements were conducted three times by adding 50 ml of synthetic medium that contained 3.5 % NaCl with 10 mg NH_4^+-N L^{-1} for AOR or 10 mg NO_2^- -N L^{-1} for NOR for the first and the third measurement. For the second measurement the ammonia concentrations were varied from 10 – 5000 mg NH_4^+-N L^{-1} (see Table 3.7). A comparison of the rates of the second and third measurement with those of the first measurement is presented in Figure 4.26.

Ammonia inhibition of AOR started at a concentration of 0.25-0.5 g N L^{-1} (Figure 4.26). The AOR reached only 50% of the initial rate when the ammonia concentration in the medium was 1 g N L^{-1}. At an ammonia concentration of 5 g N L^{-1}, the highest concentration of ammonia applied in the experiment, the AOR reached only 8%, however recovery of the AOR to 100% was achieved immediately after the biofilm was put back in the low ammonia medium.

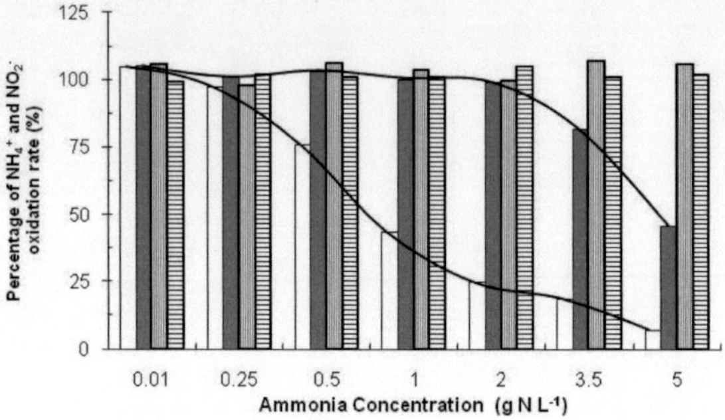

Figure 4.26 Ammonia and nitrite oxidation rates at different concentrations of ammonia by nitrifiers that formed a biofilm on porous ceramic rings. Symbols: open bars, 2nd AOR; closed bars, 2nd NOR; dotted bars, 3rd AOR; bars with horizontal lines, 3rd NOR

On the other hand, inhibition of the NOR began at ammonia concentration above 2 %, resulting in about 75 % residual activity at 3.5 g N L^{-1}. At ammonia concentration of 5 g N L^{-1}, the residual NOR was about 40%. The recovery experiment shows that the inhibition of NORs due to high ammonia concentrations was also temporary as observed for AORs.

4.11 The effect of nitrite on the ammonia and nitrite oxidation rates by biofilm-forming nitrifiers

Nitrite, an intermittent product of nitrification is known to be more toxic than ammonia for nitrification. Thus, the effect of nitrite on AORs and NORs was determined in a similar experiment as performed for determination of the effect ammonia on AORs and NORs. Instead of adding 10 mg NH_4^+-N L^{-1} for the control experiments (first and third measurement), 10 mg NO_2^-N L^{-1} was added into the reactors to determine the basic AORs and NORs. The influence of increased nitrite concentrations (10- 400 mg NO_2^-N L^{-1}) was determined in the second measurement (see Table 3.7). Results are presented in Figure 4.27.

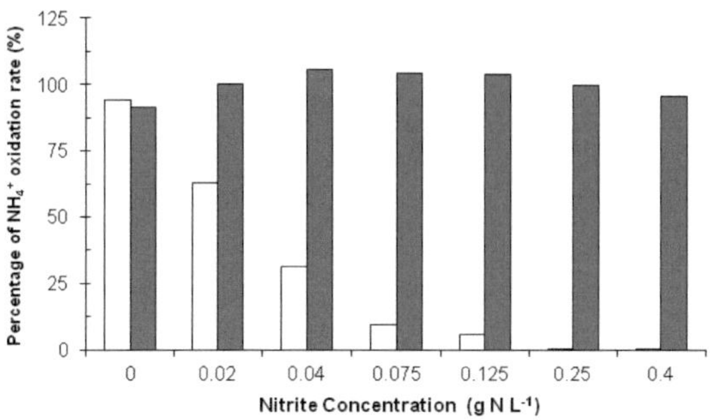

Figure 4.27 Ammonia and nitrite oxidation rates at different concentrations of nitrite by biofilm-forming nitrifiers on porous ceramic rings. Symbols: open bars, 2nd NOR measurement; closed bars, 1st and 3rd NOR measurement

The AORs decreased while the nitrite concentration increased (Figure 4.27). At a concentration of about 0.027 g N L^{-1} nitrite, the NOR was 50% compared to the initial rate and above nitrite concentrations of 0.25 g N L^{-1}, nitritation failed completely. However, the inhibition by nitrite, even at concentrations of up to 0.4 g N L^{-1} was temporary. The AOR turned back to previous rates (100%) after the biofilm was put back into a medium without nitrite. On the other hand, nitrite, even at concentrations of up to 10 g N L^{-1} did not inhibit NOR (results not shown).

4.12 The effect of nitrate on ammonia and nitrite oxidation rates by biofilm-forming nitrifiers

The procedure in this experiment was similar with the procedure used for the experiments in chapters 4.10 and 4.11. For the basic rate measurement, 50 ml of synthetic medium with 3.5 % NaCl was supplemented either with 10 mg NH_4^+-N L^{-1} or 10 mg NO_2^- -N L^{-1} and was supplied to the reactors. Varying nitrate concentrations from 0 mg N L^{-1} to 10 000 mg N L^{-1} were added after the basic measurement of the AOR and NOR for the second rate measurements. It was observed that even after applying the highest concentration of nitrate, no negative influence on AOR and NOR occurred (data now shown).

4.13 The effect of temperature on ammonia and nitrite oxidation rates by biofilm-forming nitrifiers

Bacteria involved in biological processes all grow best at their species-related optimum temperature. They might respond with lower and higher growth rates or AORs and NORs at lower or higher temperatures, respectively. To test the influence of temperature on AORs and NORs the basic AOR or NOR measurement was conducted by adding 50 ml of synthetic medium (3.5 % NaCl) with either 10 mg NH_4^+-N L^{-1} or 10 mg NO_2^-N L^{-1} to cylindrical glass reactors with nitrifying biofilm-carrying porous ceramic rings at a starting incubation temperature of 22.5 ^{0}C.

Figure 4.28 documents AORs and NORs for lower and higher incubation temperatures. As expected the incubation temperature obviously affects the rates of ammonia and nitrite oxidation.

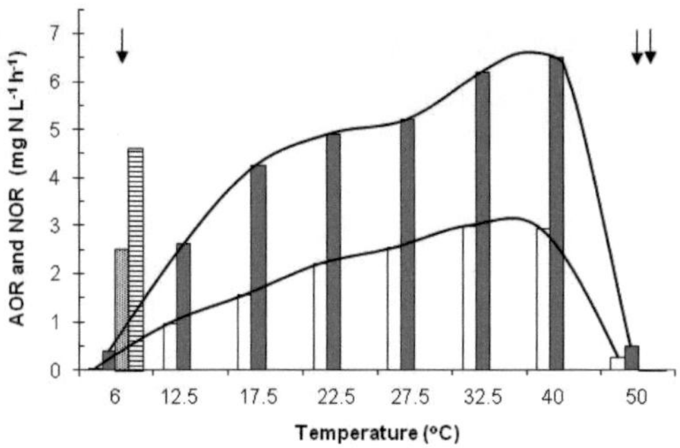

Figure 4.28 Ammonia and nitrite oxidation rates at different incubation temperature by a nitrifying biofilm on porous ceramic rings. Symbols: open bars, AOR; closed bars, NOR; (↓) recovery test at 22.5 °C after incubation at 6 °C, complete recovery at 22.5 °C; at 22.5 °C after incubation at 50 °C, no recovery (↓↓)

AORs and NORs decreased with decreasing temperatures and only a very low AOR and NOR remained at 6 °C, while AORs and NORs increased up to 40 °C with little residual activity at 50 °C. A temperature increment from 22.5 to 32.5°C resulted in an AOR and NOR increase by a factor of 1.36 (3.59% per °C) and 1.26 (2.64% per °C), respectively. An increase of the temperature from 32.5°C to 40°C did not lead to increased AORs and NORs and for higher temperatures than 40 °C AORs and NORs drastically decreased. This indicated that the optimal temperature for nitrification was lower than 40 °C. The AORs and NORs decreased to 0.44 (5.59% per °C) and 0.53 (4.67% per °C), respectively, when the incubation temperature was decreased from 22.5 to 12.5°C. At 6°C, no oxidation activity of AOB remained, while the residual activity of NOB was only about 10%.

Nitrifiers recovered when the incubation temperature was raised from 6 °C back to 22.5 ^{0}C (↓) but no recovery was possible after an incubation at 50 ^{0}C (↓↓), showing that the effect of the low temperature was temporary. In contrary, the effect of the high temperature was more severe and no immediate recovery was observed.

4.14 Effect of aerobic and anaerobic storage of nitrifiers on porous ceramic rings on AOR and NOR

Nitrifying bacteria are characterized as slow-growing autotrophic bacteria. In practice, relatively long start-up phases are needed for a nitrification reactors to function properly. Storage of surplus biomass as reserve inocula would be a means to shorten the start-up phase of new reactors, after failure or after closure for maintenance.

To test the survival of nitrifiers in the reactor effluent 25 ml of liquid from a CSTR was aerobically stored in 100 ml serum flasks. After 1, 4, 6 and 25 days storage time AORs and NORs were determined by adding 10 mg NH_4^+-N L^{-1} or 10 mg NO_2^--N L^{-1} into respective flasks.

As obvious from Figure 4.29, the AOR seems to decrease earlier than the NOR with increasing storage time under anaerobic conditions. After 25 days of storage, AOR and NOR were 60 % and 80 % of initial rates, respectively, indicating a slightly higher loss of activity of the AOB than the NOB.

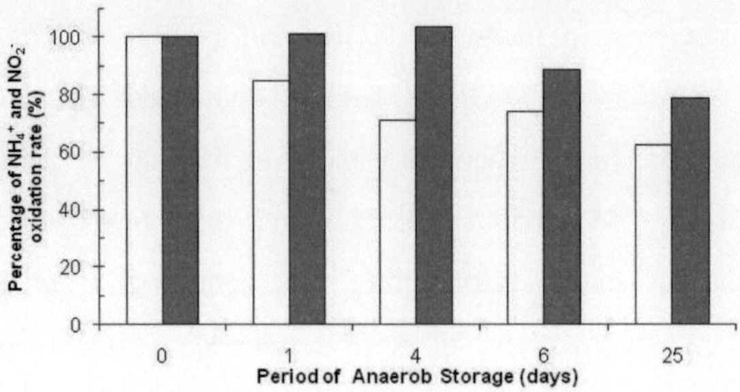

Figure 4.29 Ammonia and nitrite oxidation rates of suspended bacteria after a period of storage under the exclusion of oxygen. Symbols: open bars, AOR; closed bars, NOR after oxygen supply

The effect of storage in fixed bed reactors (FBR A and FBR B) was also examined. At steady state conditions with no ammonia and nitrite in the effluent, the reactors were operated batch wise and then AOR and NOR was measured by adding substrate solution up to 60 mg NH_4^+-N L^{-1} or 20 mg NO_2^--N L^{-1}. After the rate measurement was finished, for the three following days, the reactors were still operated in batch mode without adding any ammonia or nitrite. During the three days, the reactors were aerated to simulate aerobic storage and then were not aerated for the anaerobic storage. Then rate measurements were conducted and their result are presented in Figure 4.30.

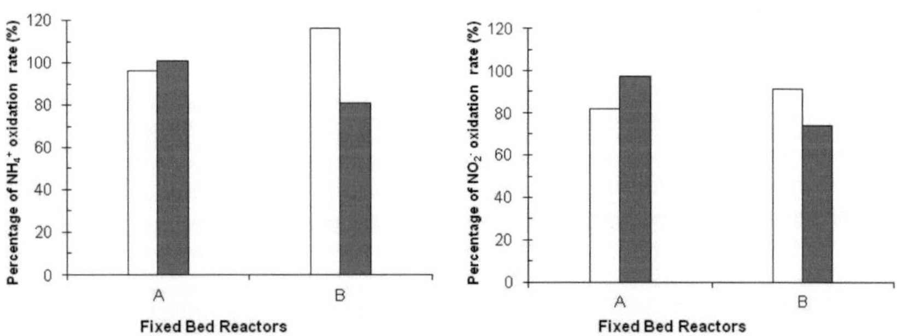

(a) (b)

Figure 4.30 AOR (a) and NOR (b) of a nitrifying biofilm-forming biomass in FBR A and B after 3 days of aerobic storage (open bars) and 3 days of anaerobic storage (closed bar)

Neither aerobic nor anaerobic storage influenced AOR in FBR A. In FBR B, three days of aerobic starvation seemed to increase AOR slightly, whereas the anaerobic starvation seemed decreased AOR slightly.

Based on Figure 4.30, it can be said that NORs in FBR A were more influenced by aerobic storage (80%) than by anaerobic storage (100%). In FBR B anaerobic starvation decreased NORs to75%, whereas starvation under aerobic conditions decreased NORs only to 90%.

4.15 Kinetics of ammonia and nitrite oxidation at different starting concentration

In this experiment, FBR A and B were operated batch wise and then 10 mg NH_4^+-N L^- or 15 mg NO_2^--N L^{-1} was added into the reactors. Samples were taken from the reactors every 10 minute to obtain in detail the trend of ammonia or nitrite oxidation. The results of the analyses are presented in Figure 4.31.

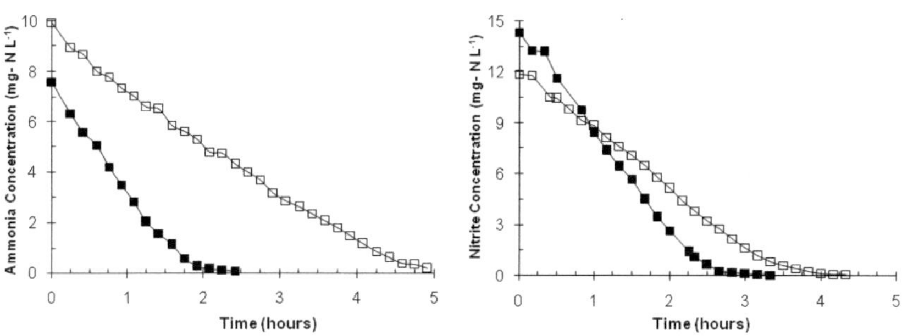

(a) (b)

Figure 4.31 Ammonia (a) and nitrite concentrations (b) during nitrification tests in FBR A (closed square, ■) and FBR B (open square, □)

In both FBRs, nitritation rates (AOR) were higher than nitratation rates (NOR), which was expected for balanced nitrification. In FBR A, at a nitroten concentration of above 1.5 mg N L^{-1}, the biofilm could oxidize relatively uniform ammonia and nitrite. Much lower AORs and NORs were obtained at lower concentration of ammonia and nitrite (under 1.5 mg N L^{-1})

Chapter 5

DISCUSSION

5.1 Microbial analysis

- *Selection of halophilic enrichment cultures*

An often reported but not always successful approach for biological nitrification of saline wastewater is the stepwise adaption of non-saline water bacteria to halophilic conditions (e.g. Moussa et al. 2006). Adaption of a sweat water nitrifying population to saltwater conditions of up to 40 g L^{-1} NaCl seems not to be possible (Moussa et al. 2006), whereas an enrichment of halophilic nitrifiers from halophilic environmental sources would be a more promising approach, since most halophilic nitrifiers are apparently ubiquitous in seawater environment (Francis et al. 2005).

Enrichment of halophilic nitrifying consortia from seawater/mud samples of the coastal region of the North Sea in Germany with a salinity of about 3% NaCl (conductivity > 30 mS cm^{-1}; Table 4.1; Figure 3.7) seemed thus to be a suitable approach to establish halophilic nitrification in the laboratory (Sudarno et al. 2010). The applicability of marine sediments as inocula for nitrification under saline conditions was also reported by Antileo et al. (2002), who found ammonia oxidation after a lag-phase of 15 days. In our first enrichments from different marine sources, AORs ranged between 4.9 and 15.1 mg N L^{-1} day^{-1}, which was in accordance with the work of Rejish Kumar et al. (2009), who measured AORs of 4.6 and 12.2 mg N L^{-1} day^{-1}.

- *Biofilm growth*

The bacteria involved in nitrification are autotrophic microorganisms with carbon dioxide serving as their carbon source and have slow growth rates. An even slower growth rate of nitrifiers apparently is observed for nitrifiers in saline environment. A quantitative and qualitative estimation of their specific concentrations, especially of "biofilm nitrifiers" is a very difficult task (Lazavora and Manem 1995) and requires molecular biological methods.

The development of bacterial growth can be examined by measuring bacterial biomass, metabolic activity and by counting cells. This examination will be easy in suspended cultures but is much more difficult for biofilm-forming bacteria. Detachment of immobilized bacteria is very difficult and incomplete, especially during the first growth phases where the biofilm is still faint and not distributed uniformly on and in the pores of the substratum. Later on during biofilm development not only growth but also other processes, such as decay and detachment play a significant role.

Due to the expected inaccuracy of population density measures, quantitative and qualitative estimation of biofilm specific parameters were conducted by measuring nitrification rates and by comparing ammonia and nitrite oxidation.

AOB attached faster than NOB in FBR A and FBR B as shown from ammonia and nitrite oxidation rates until days 50 (Figure 4.17b, Table 4.2). Furthermore, ammonia oxidizer recovered faster than nitrite oxidizers after biofilm detachment in control reactor FBR I (Figure 4.10). The domination of AOB was also reported on the surface of the BioCube sponge media (Chae et al. 2008). The initially faster attachment of AOB on surfaces for biofilm formation indicates that AOB have a better capability to attach to mineral or plastic surfaces or grew faster than NOB. Except for that, it is possible that NOB will be active after nitrite was produced by AOB, as perfectly explained by Peng and Zhu (2006). NOB will exist after AOB exist, and later on during the nitrification process, AOB and NOB co-exist and benefit from the close physical association. On one side, the close physical association is useful for energetic reasons. NOB are able to efficiently intercept the nitrite produced by the AOB, helping to cope with the poor energy yield of nitrite oxidation. On the other side, AOB are dependent on the presence of the NOB, as the latter relieve them from the toxic nitrite. It therefore aids in the defense against the toxicity of nitrite by preventing its accumulation or formation of toxic by-products such as NO that can inhibit bacterial enzymes.

During the period from day 0 to day 50, growth of AOB was not stagnant as judged from ammonia oxidation rates, while NOB seemed to develop not as fast. A population shift from AOB to NOB apparently occurred only upon further incubation after day 50 (Figure 4.17b). This was similarly the case in FBR A and FBR B (Table 4.2), in which the reversal of oxidation rates happened after days 84 for FBR A and day 64 FBR B. However, the AOR was more instable compared to NOR.

Biofilm growth and maturation is characterized by formation of strong micro colonies which are very resistant against high shear forces and different physical/chemical manipulations. Biofilms grow during the developmental phase and biomass increases whereas after reaching an equilibrium in maturated biofilms, microbial decay processes in deeper layers, that are no longer optimally provided with oxygen start, ammonia or nitrite enrichment by incomplete oxidation may take place and may finally lead to sloughing off.

The maturation state was examined from the beginning in the halophilic control reactor (FBR I, Figure 4.10; Table 3.3). The growth rates in the reactor only increased by a factor of 1.5 for AOB after 99 days and 1.1 for NOB after 70 days. The value is much lower than that reported by Bock et al. (cited by Peng and Zhu, 2006). They reported that generation times for AOB and NOB were 7–8 hours and 10–13 hours, respectively. Total biofilm mass during the experiment, then, could be said was constant and a maturation state was already reached.

The maturation state was also obtained in the non-halophilic control reactor FBR N (Figure 4.13, Table 3.4). After an acclimatization period for two weeks by batch wise operation, AOR and NOR steadily increased (with an increasing factor of 1.3 and 1.2) during the experiment indicating a slow increase of the biofilm mass. Fresh polyethylene/clay was applied in FBR N as supporting material, so the relatively fast maturation in the reactor were likely made possible by two weeks batch wise pre-incubation. Characklis and Marshall (1990) found that the maturation process was achieved after a steady state period of a few days already.

A permanently decreasing NOR (FBR I, Figure 4.10) might be due to a spontaneous detachment. Dalton et al. (1996) observed a spontaneous detachment after maturation of a biofilm. They attributed the spontaneous detachment of cells from the biofilm to two processes, erosion and sloughing, based on the magnitude and frequency of the detachment events. Erosion is the continual detachment of single cells and small portions of the biofilm, whereas sloughing off is the rapid, massive loss of parts of the surface biofilm layers due to gravity. On the other hand Elenter et al. (2007) studied a biofilm of autotrophic bacteria and obtained a significant amount of lost biofilm during constant shear stress conditions. In our experiment, the spontaneous detachment could be categorized as sloughing off as judged from the detached material.

Surprisingly the "sloughing off" seemed to effect more the NOB and not so much the AOB, as judged from activities (Figure 4.10). This is difficult to explain, since most studies have shown that NOB usually are located in the depth of the biofilm below an external AOB layer. AOB and NOB in a biofilm exist as micro colonies: if sloughing off of NOB biofilm layers occurs this automatically leads also to sloughing off of AOB.

A permanent decrease of the NOR seemed to occur in the control reactor with 3.5% salinity (Figure 4.10), but not in the control reactor with 0.03% salinity (Figure 4.13). This fact is not a strong indication that salinity deteriorates nitratation activity in a biofilm. In other reactors (FBR F, G, H, Figure 4.7; FBR J, K, L, M, Figure 4.13), no permanent declining of NOR was obtained, even if fluctuating salinity was applied to the reactors.

- *Microbial population*

Molecular biological analysis applying the Polymerase chain reaction (PCR) in FBR B revealed that the biomass in the reactors was dominated by *Nitrosomonas* sp. and *Nitrospira* sp. (Sudarno et al. 2010). The 16S rRNA gene sequences for AOB in FBR B detected *Nitrosomonas aestuarii* and *Nitrosomonas* Nm143, whereas the 16S rRNA gene sequences for NOB detected *Nitrospira marina* as the dominant nitrifying bacteria. Those sequences correlate with the sequences, which are usually found from coastal and marine water or sediments (Watson et al. 1986, Koops et al. 1991, Purkhold et al. 2003). Due to an increasing ALR and a decreasing pH in FBR B, a shift within the ammonia oxidizing population in the genus *Nitrosomonas* at the subspecies level occurred, whereas the nitrite oxidizing *Nitrospira* population apparently did not change (Sudarno et al. 2010).

5.2 Feasibility of fixed bed reactors for nitrification of saline wastewater

- *Supporting material*

FBR A with polyethylene/clay sinter lamellas and FBR B with porous ceramic rings as substrata could be operated for a long period of time under high salinities and ammonia concentrations. The biofilm may have been formed initially by fast-growing heterotrophic bacteria and during a much longer time of operation under autotrophic conditions by the slow-growing autotrophic nitrifying bacteria. For initial attachment surface characteristics of the support material are important, e.g., the roughness of surfaces or other surfaces properties such as surface charges, hydrophobicity or hydrophilicity and significantly influence bacterial colonization (Verran et al. 1991, Gjaltema et al. 1997). Crevices and pores act as niches with moderate turbulences of the medium for better attachment of bacteria and protect the biofilm from shear forces (Fox et al. 1990).

From the beginning of the continuous operation it could be observed that FBR B had a better nitrification performance (Figures 4.3 and 4.4). In FBR B, porous ceramic rings with a specific surface area of 934 m^2 m^{-3} were used as support material for biofilm formation, which was a twice as high specific surface area than was available on the polyethylene/clay sinter lamellas of FBR A.

Ammonia oxidation in FBR B was almost complete except when titration failed. Nitrite accumulation was only seen during stage 3 (Table 3.1) after re-installation of the pH titrator (Figure 4.4c). This was in accordance with Krüner and Rosenthal (1983) who reported that the AOR was proportional to the surface area of the support material that was used for biofiltration. It might be generalized that the surface area of the support material significantly influences the conversion rates when only a faint biofilm, as in the case of autotrophic nitrifiers, was formed, whereas the inner surfaces of pores play only a minor role if a thick biofilm was obtained as for carbon-rich industrial wastewater. Inner surfaces are then blocked and diffusion of substrates into pores is highly limited.

The maximum AORs and NORs before installation of the liquid recirculation were 6 and 7 mg N L^{-1} h^{-1}, respectively in FBR A and 5 and 5.9 mg N L^{-1} h^{-1} in FBR B. To correlate the specific surface area of the support material with the maximum N-removal rate, area-related AORs and NORs were calculated. For FBR A, a maximum surface area-related AOR and NOR of 312 and 386 mg N m^{-2} day^{-1} and for FBR B of 199 and 236 mg N L^{-1} day^{-1}, respectively was determined (Table 4.2). Maximal and area-related AORs and NORs were better in FBR A than in FBR B, but the overall performance (residual ammonia in reactor effluent, intermediate nitrite accumulation) was better in FBR B. This may be due to a surface-oriented and more dense biofilm in FBR A with higher rates but less efficient conversion. Nijhof and Bovendeur (1990) determined an area-related AOR under saline conditions at 24°C of 280 mg N m^{-2} day^{-1}, whereas a much better AOR under freshwater conditions of 690 mg N m^{-2} day^{-1} was obtained.

- *Influence of pH*

During the continuous operation of FBRs A and B for nitrification under halophilic conditions, interferences related to pH, alkalinity and mixing could not be avoided, which may led to adverse effects. Failure of pH titration (Figure 4.3, days 10, 38, 67, 87 and Figure 4.4, days 10, 45) caused incomplete ammonia oxidation in FBRs A and B. Precipitation of inorganic compounds at the diaphragm of the pH electrode might been the reason for the very short stable operation time leading to the addition of too much acid or caustic. Failure leading to a high pH was more adverse than failure leading to a low pH. High pH caused incomplete ammonia oxidation and in addition nitrite accumulation (Figure 4.4, day 10), while low pH (Figure 4.4, day 45) only lead to higher ammonia concentrations in the effluent. The pH effect on nitrification was only temporary; full ammonia oxidation was obtained after exchange of the pH probe within the next 10 days. Siegrist and Gujer (1987) for instance had reported that the ammonia oxidation rate recovered after exposure to a lower pH.

Mechanisms of pH inhibition of nitrification might be an activation–deactivation of nitrifying bacteria, nutritional effects connected with alkalinity or inhibition through free ammonia and free nitrous acid, as summarized by Villaverde et al. (1997). In this study deactivation of nitrifiers might likely be caused by the pH inhibition mechanism. The availability of a carbon source for bacterial growth was ascertained by application of $NaHCO_3$ as pH buffer and the calculated FA and FNA, based on Equation 2.12 and 2.13, were lower than inhibition causing concentrations.

- *Influence of pH, alkalinity and ammonia loading*

Villaverde et al. (1997) described a linear correlation between alkalinity (e.g. expressed as milligrams per liter $CaCO_3$) and pH with a stoichiometric coefficient of 7.1 mg $CaCO_3$ consumed per milligrams NH_4^+-N oxidized. If the alkalinity was not sufficient, the pH decreased during oxidation of 50 mg N L^{-1} ammonia as it was observed in our original samples A and C and to a minor extent in sample D to values below pH 5 (Figure 4.1). Sample B, a seawater/mud mixture, had the highest alkalinity, and the pH was stable above 7.5 during the whole experiment, including a second feeding with 50 mg N L^{-1} ammonia (Figure 4.1b)

In FBR A and FBR B, maintaining the system of the reactors in an optimum pH range of 8 $\pm$ 0.2 was conducted by a pH titrator with 0.25 M KOH solution (Table 3.1) or by addition of $NaHCO_3$ as pH buffer (Table 3.1, phase d). A pH titrator provided addition of acid or base in accordance with the required exact amount. When the pH in the medium was lower or higher than the desired value, acid or base solution was immediately pumped into the medium. Correct titration was dependent on the function of the pH electrode. A malfunction of the pH electrode due to a blockage of the diaphragm may lead to a very low or high pH value in the FBRs and to failure of nitrification.

The use of $NaHCO_3$ was chosen as a pH buffer system and also as a source of carbon for bacteria growth. When the concentration of ammonia in wastewater increased, more $NaHCO_3$ must be added to stabilize the pH. Controlling of the pH with $NaHCO_3$ at the lab scale is usually done by mixing an appropriate amount of $NaHCO_3$ into the medium (7.1 gram alkalinity as $CaCO_3$ or 12 g as $NaHCO_3$ for 1 gram ammonia in the medium). In the field scale this solution is fed right before the nitrification plant.

The alkalinity: ammonia ratio (7.1:1) is based on an assumption that all ammonia will be oxidized. $NaHCO_3$, therefore, will be excessive, if nitrifiers oxidize ammonia incompletely. The surplus $NaHCO_3$ could be avoided by addition of the solution with a pH titrator. Carrera et al. (2003) reported that as biomass ability for oxidizing ammonia changes, some problems related to alkalinity addition appeared. These problems were mainly due to the fact that it was difficult to dissolve sodium bicarbonate in the wastewater completely at the targeted concentrations.

Initially, the pH could be successfully maintained at about 8 with the addition of $NaHCO_3$ buffer (Figures 4.3 and 4.4, days 187–194). However, white crystals over supporting material or at the walls of the inlet tube were observed during the use of this buffer. This led to a disturbance of the high rate reactor performance. A similar phenomenon was also reported by Vredenbregt (1997) and his group, where the level of scaling due to calcium carbonate formation in the reactor was so high that the reactor had to be shut down. Sakairi et al. (1997) reported that bicarbonate ions were deprotonated in the presence of ammonia. This raises the carbonate concentration in the solution to such an extent as to exceed the solubility product of calcium and magnesium carbonates (Ca^+ and Mg^{2+} are present in seawater), which then precipitate out of the solution.

- *Ammonia loading*

The ammonia loading rate is the amount of ammonia as nitrogen equivalents that are added per liter of reactor volume per day. In FBRs A and B, ALRs were increased two times by increasing the ammonia concentration in the influent or shortening the HRT (Table 3.1). Increase of the ALR by reducing the HRT (Figures 4.3b and 4.4b, days 56–155) led to an increase of the ammonia in the effluent only in FBR B. A further increase of the ALR by increasing the ammonia concentration in the medium (Figures 4.3c and 4.4c, days 156–215) resulted also in an incomplete ammonia oxidation in effluent of FBR A.

When the HRT was shortened, immobilized bacteria had insufficient contact time to convert the substrates. Apart of that, the vertical flow of medium increased, leading to a higher turbulence in the reactor. The higher turbulence then affected the thickness of the biofilm on the substrata and subsequently the diffusion of substrate from the bulk liquid into the biofilm (Seo et al. 2001).

Nitrifiers might oxidize not all of the ammonia, when the ALR was increased by elevating ammonia concentrations. Higher remaining ammonia concentrations and intermediate product concentrations (nitrite) in the medium due to restricted ammonia oxidation could bring about inhibition for nitrifiers and finally disturbance of both nitrification reactions.

An effect of the ALR by increasing ammonia concentration can be obviously observed in FBRs O and P (Figure 4.15). When the ALR in FBR O was doubled (Figure 4.15, phase III), at steady state during continuous operation no ammonia and nitrite was obtained in the effluent as in the previous phase (phase II). The oxidation rate measurements indicated higher AOR and NOR (Figure 4.16) and thus indicating that the maximum loading was not yet reached. After the ammonia concentration in FBR 0 was increased from 85 to 125 mg N L^{-1} (Figure 4.15, phase IV), the AOB in the continuously run reactor could only oxidize half (62.5 mg N L^{-1}) of total ammonia supplied with the influent and the NOB consumed almost none of the nitrite, which was produced by AOB. In phase III, AOB and NOB could oxidize 85 mg N L^{-1} of ammonia and nitrite, respectively. This means that the capability of both sorts of nitrifying bacteria decreased due to the increased ALR.

When the ammonia concentration in FBR P was increased from 40 to 250 mg N L^{-1} (Figure 4.15, phase III), ammonia in the effluent immediately increased to 150 mg N L^{-1} and remained constant until day 115. The nitrite concentration increased steadily to about 50 mg N L^{-1} at day 115. This might indicate that an ammonia concentration of 150 mg N L^{-1} could inhibit NOB but not AOB under these incubation conditions. Furthermore, increasing nitrite accumulation might be also caused by inhibition of NOB through nitrite. At day 115 nitrite accumulation was 100 mg N L^{-1}. From this day onward, ammonia concentrations in effluent tended to increase, indicating that AOB were inhibited by nitrite concentration of 100 mg N L^{-1}. Inhibition of the nitrifiers at stepwise increasing ALRs was found, but with suitable operational strategies, such as minimizing the hydraulic load or dilution of the wastewater at the beginning of the change, the inhibition could be limited or even avoided.

- *Influence of shear forces*

Shear stress could cause biofilm detachment. When the shear forces exceed the biofilm mechanical strength, detachment occurs. In FBRs A and B, biomass detachment was observed at increased shear forces immediately after installation of liquid recirculation. Concomitant with this phenomenon, AORs decreased notably, while the NORs were apparently not influenced (Figures 4.3d and 4.4d). The decrease was more pronounced in FBR B with porous ceramic rings as a carrier material than in FBR A with polyethylene/clay sinter lamellas. Turbulence in this reactor was visibly higher than in FBR A due to more rugged surfaces through which the water must find it way to the top. Also, some loose sediments on surfaces that were not directly in the up-flow path of the wastewater may have been swept away at higher upflow velocity.

This might never the less indicate that the ammonia oxidizing nitrifiers were mainly located in the outer layers of the biofilm on the support material and therefore may have been sheared off to a higher extent by increased shear forces during liquid recirculation. This would be in accordance with the report of Okabe et al. (1999), who found a layering of AOB at the surface and of nitrite oxidizing bacteria in deeper zones in nitrifying biofilms fed with domestic wastewater.

Shear stress could also induce mass transfer changes. A short-term effect immediately after installation of the liquid recirculation was an increase of the AOR and NOR in FBR A (Table 4.2, day 217) indicating that the better mixing, created by aeration and an increased up flow velocity (1.25 m h^{-1}), apparently improved mass transfer of oxygen and ammonia from the bulk liquid into the biofilm. Zhu and Chen (2001) also observed an influence of turbulence on the AOR. The performance of their nitrifying biofilters could be significantly improved by increasing the Reynolds number in their biofilm reactor. Therefore, there may be a way to improve the nitrification efficiency of a fixed film biofilter through turbulent flow. In FBR B, the aeration system was apparently sufficient for an optimal supply of the nitrifiers in the biofilm with oxygen and ammonia since the AOR and the NOR did not increase after starting the liquid recirculation (Table 4.2, after day 215).

In FBR B, porous ceramic rings were not arranged uniformly in the reactor, leading to a non uniform distribution of liquid flow and subsequently higher shear stress. Biofilm detachment increased with increasing turbulence (Vieira and Melo 1999). Lazyrova et al. (1994) reported that a very thin biofilm could be maintained at a high turbulence. In Figure 4.18b, a thick but not compact biofilm was located inside and at the top of the rings, where shear stress was low and detachment was minimized.

In other areas of the porous ceramic rings, biofilm was invisible with bare eyes, but could be seen by scanning microscopy. An invisible, faint biofilm was obtained on new porous ceramic rings after incubation for 50 days in FBR B (see section 3.3.2.2). Measurement of AOR and NOR with porous ceramic rings that carry an invisible biofilm verified the existence of the nitrifiers (Figure 4.17b).

- *Storage of biofilm-carrying substrata*

Complete shut-down of operation of wastewater plants for a few days or even months might be necessary for maintenance or up-grading of the plant. During this time it is important to maintain the viability of biomass during more or less long starvation periods. If the biomass after storage would not function properly due to permanent inactivity of bacteria, a complete new start-up with a new substratum must be carried out. Active biofilm-carrying substrata, that are stored at proper conditions in a storage facility, should be kept available to shorten the start-up period. Biofim nitrifiers in FBRs A and B still were fully active after an aerobic and anaerobic starvation period of 3 days (Figure 4.30). Furthermore, it was possible to store suspended nitrifiers up to even 25 days with only slight decrease of activities (Figure 4.29).

Dahl et al. (1997) investigated nitrification at high salinity of 20 g Cl^{-1} L^{-1} and reported that anaerobic storage of activated sludge during up to four days at 20–23°C only slightly affected the nitrification activity, as approx. 80% of the initial activity was still present after four days. Also, it is possible to store excess sludge anaerobically up to four days to serve as backup sludge for restarting the plant. Whereas Salem et al. (2005) found that the decay rate of ammonia oxidizers after 14 days of starvation at 20°C was 0.02 day^{-1}, Furukawa et al. (1993), who stored the nitrifiers in a refrigerator for 6 days, reported that the pellets of acclimated marine nitrifying sludge could regain their full activity.

Furthermore, Salem et al. (2005) stated that the decay rate of nitrifiers might have been underestimated, because of the application of a very rapid test method: The behavior (nitrifying activity) of the bacteria was tested immediately after adding ammonia to a starved culture. This procedure does not consider lag-time for enzyme reactivation and thus tends to overestimate decay rates. Some decay rates of nitrifiers at different condition and temperatures that were reported in the literature are summarized in Table 5.1.

Table 5.1 Decay rate of the nitrifiers at different condition and temperatures

Nitrifiers	Storage			decay rate (per day)		Literature
	Condition	Temperature (°C)	Duration (day)	Ammonia oxidizers	Nitrite oxidizers	
Immobilized		4	6	0.00	-	Furukawa et al. 1986
Biofilm	anaerobic	20	4	0.05	-	Dahl et al. 1997
Suspension	Aerobic	20	14	0.02	0.08	Salem et al. 2005
Suspension	Anaerobic	20	14	0.08	0.07	Salem et al. 2005
Granuler	aerobic	4	90	0.012	-	
			60	-	0.024	Wang et al. 2008
Biofilm	Aerobic	22.5	3	0.03	0.045	This study
Biofilm	Anaerobic	22.5	3	-	0.048	This study
Suspension	Anaerobic	22.5	25	0.016	0.008	This study

5.3 Effect of salinity changes on nitrification

- *Porous ceramic rings containing FBRs (Figure 4.5, FBRs C, D, E) compared to Polyethylene/clay sinter lamellas containing FBRs (Figure 4.8, FBRs F, G, H, I) with a starting NaCl concentration of 3.5%*

Varying wastewater composition and salinity in industrial processes are the main factors complicating the application of nitrification of saline wastewater. The declining rate of ammonia and nitrite concentrations in range of 20–5 mg N L^{-1} and 15–2.5 mg N L^{-1}, respectively, at different salt concentrations was the same (Figure 4.6). A more detailed measurement of oxidation rates by biofilm nitrifiers also revealed a similar trend (Figures 4.31a and b). The ammonia and nitrite oxidation rates, therefore, were determined based on the slope of the concentration decline.

When the salinity was reduced from 3.5 % to 0.03%, no ammonia was detected in effluent from FBRs F, G and H (Figure 4.8, phase IV) until 5 days of incubation time, whereas ammonia was immediately obtained in FBRs C, D and E (Figure 4.5, phase III). Considering that the reactors were initially inoculated by sludge taken from the same location, the different effect might be attributed to density and thickness of the biofilm.

Ammonia oxidation was inhibited the more, the higher the salinity was increased. Significant nitrite accumulation during the higher salinity was observed only in FBRs F, G, H (Figure 4.8, phase VI) but not in FBRs C, D and E (Figure 4.5, phase V).

The salinity increase by a factor of 2.5 (from 3.5 to 9%) in FBR A led to a severe decrease of the ammonia removal efficiency to only residual 15%, whereas a salinity increase by a factor of 2 (3.5 to 7%, FBR B) or 1.5 (3.5 to 5%, FBR C) reduced the ammonia removal efficiency less drastically to 51% or 67%, respectively, without significant nitrite accumulation (Figure 4.5, Table 5.2). These results are in accordance with those observed by Dahl et al. (1997) and Uygur (2006), who found a decrease of the ammonia oxidation activity in the range of 30% after increasing the salinity from 3.5 to 5% or of 40% after increasing the salinity from 0 to 6%.

Table 5.2 Effects of salinity change on several parameters

Parameter	FBR									
	C	D	E	F	G	H	I	J	K	L
Sal. changes (%)	3.5 to 9	3.5 to 7	3.5 to 5	3.5 to 9	3.5 to 7	3.5 to 5	0.03 to 10.5	0.03 to 9	0.03 to 7	0.03 to 5
Efficiency (%)[a]	15	51	67	75	85	93	0	75	100	100
AOR (%)[b]	17	44	62	55	61	75	0	13	70	74
NOR (%)[b]	55	66	110	38	46	65	0	0	7	16
MNA[c] mg N L^{-1}	0	4[d]	3[d]	21	15	13	0	31	73	47
NOR:AOR	6	2.2	2.08	1.5	1.4	2.7	0	0	0.15	0.25

[a] = In all reactors efficiency before changing salinity was 100%
[b] = Percentage AOR or NOR after changing salinity compared to original rate.
[c] = Maximum nitrite accumulation
[d] = Found only in the initial phase.

The salinity increase in FBRs F, G and H by a factor of 2.5, 2 and 1.5, respectively, only resulted in a slightly decrease of the ammonia removal efficiency (Table 5.2). The real inhibiting effect could not be seen clearly, because the ammonia oxidizers in the FBRs might have had the ability to oxidize more ammonia than the concentration that was added (120 mg N L^{-1}).

The fast and complete recovery of FBR C might have been caused by precipitation of a white amorphous material, presumably $CaCO_3$, which stabilized the pH or maintained a better supply of carbon dioxide for nitrifiers. No such precipitates were observed in FBRs D and E.

- *Nitrification rates in FBRs C, D, E (Figure 4.7) compared to those in FBRs F, G, H (Figure 4.9)*

Under batch incubation, in the first phase of FBRs C, D and E, ammonia oxidized per day was 48 mg N L^{-1} (AOR = 2.2 mg N L^{-1} h^{-1}, Figure 4.7). No ammonia, however, was detected in effluents during continuous operation (Figure 4.5). It seemed that a better ammonia oxidation was observed during continuous than batch incubation.

A higher decrease of the AOR due to the salinity increase by a factor of 2.5, 2 and 1.5 was found in FBRs C, D and E compared to FBRs F, G and H (Table 5.2). The explanation for the difference might not be addressed to nitrite accumulation during continuous operation, since nitrite accumulation was higher in FBRs F, G and H (Table 5.2). This is supported from the previous discussion (see section 5.2, part ALR), in which it was explained that inhibition initiating nitrite concentrations for ammonia nitrifiers were 150 mg N L^{-1}.

In contrary, the increasing salinity influenced more severely the NOR in FBRs F, G and H than in FBRs C, D and E. The combination of high salinity and nitrite accumulation during continous incubation of the reactors may be the reason for that severe effect.

The low ammonia oxidation rates in FBRs D and E at a HRT of 1 day and a substrate concentration of 60 mg L^{-1} NH_4^+-N in phase VI apparently led to residual ammonia in the effluents (Figure 4.5, phase VI). Using the AORs (FBR D = 1.50 , FBR E = 1.66 mg N L^{-1} h^{-1}, Figure 4.7) that were determined during batch incubation 36 to 40 mg L^{-1} NH_4^+-N could be oxidized by the nitrifiers and thus 24 or 20 mg L^{-1} NH_4^+-N should maximally remain. The actually remaining ammonia concentrations were somewhat lower than those, which were calculated from the AORs from batch incubations, indicating that under continuous incubation conditions the ability to oxidize ammonia might have been a little better.

In FBRs C, D and E, the NOR was similar before and after the phase of high salinity. In spite of a higher steady state NOR than AOR at every salinity (Figure 4.7), some accumulation of nitrite was detected in the first days after almost every change of salinities (Figure 4.5). A higher sensitivity of NOB to osmotic stress, as caused by changing salinities and subsequently reduced NORs was also reported by Jin et al. (2007). This indicated a high sensitivity of NOB to immediate changes of the salinity but also a high capability to adapt or recover fast. Similar results were obtained for salt-adapted nitrifiers (Dahl et al. 1997).

A fast recovery of oxidizing capability was also obtained in FBR F, G and H after the high salinity phase (Figure 4.9). Nitrite was detected at the end of the higher salinity phase in FBRs F (22 mg N L^{-1}) and G (15 mg N L^{-1}), before AOR and NOR was determined (Figure 4.8, phase VI). NOR:AOR ratios were higher than 1 (Table 5.2).

AORs of FBRs C, D, E, F, G and H were less sensitive and similar for salinities of 0.5 to 3.5% (Figures 4.5 and 4.7). At steady state conditions of FBRs C, D and E a minimum NOR:AOR ratio of 1.22 and a maximum NOR:AOR ratio of 6.0 was determined, which should not lead to nitrite accumulation. Only if the NOR:AOR ratio was always higher than 1, nitrite could not accumulate (Kim et al. 2008). Thus, accumulation of nitrite in the continuous FBRs during changes of the salinity indicated a higher sensitivity of NOB than of AOB for varying reaction conditions.

During phase V ammonia oxidation was most affected in FBR C at 9% salinity (Figure 4.5). The AOR corresponded to only 9 mg L^{-1} d^{-1} and the NOR to > 9 mg L^{-1} d^{-1}, since no nitrite was detected. During the 14 days of highly increased salinities in FBR A (9%) and also in FBRs B and C (7 and 5%, respectively) in phase V at low AOR and NOR and thus at low energy supply AOB and NOB apparently survived. Decay rates of 0.08 day^{-1} for NOB and 0.02 day^{-1} for AOB were reported by Salem et al. (2005) at 20°C, aerobic conditions and a starvation phase of 14 days.

- *Polyethylene/clay FBRs with starting concentration of 0.03% salt (FBRs J, K, L, M and N)*

A higher short-term sensitivity of NOB than of AOB and a fast recovery of NOB was also found in polyehylene/clay FBRs with starting concentration of 0.03% (FBRs I, J, K and L). After the salinity increased from 0.03% to 3.5% nitrite was only detected in the beginning of the phase, especially in FBRs L and

M (Figure 4.11, phase III). NORs in both reactors were lower than AORs during batch incubation (Figure 4.12a, b).

The higher vulnerability of NOB than AOB provides a condition in which ammonia oxidation (nitritation) proceeds, but not nitrite oxidation (nitration) occurs and consequently no nitrate was found in the medium. Such a condition is a requirement of partial nitrification, as it occurred in FBR K in phase V, after salinity was increased from 0.03% to 9% (Figures 4.11 and 4.12a, b). In FBR C (Figure 4.5) and FBR F (Figure 4.8) during the phase where the reactors were also fed with 9% of salinity, no nitrite was accumulated. A different bacterial population might be a reason. FBRs C and F were inoculated with nitrifiers from marine water, whereas FBR K was inoculated with nitrifiers from fresh water.

Nitrite accumulation at the end of periods with increased salinities was only observed during phase V after the increase of salinity in FBRs K, L, M and also during phase VI for FBRs J and L. The NOR:AOR ratios during those phases were always lower than 1 (Table 5.2). Kim et al. (2008), who observed nitrification in batch assays inoculated with nitrifiers from a wastewater plant with low salinity, reported that nitrite accumulation occurred when the NOR:AOR ratio was lower than 1. In this study, the statement of Kim et al (2008) was only valid for FBRs with starting concentrations of NaCl of 0.03%.

- *Batch assays using immobilized nitrifiers in porous ceramic rings (3.5% salt)*

In fact, the respond of AOR and NOR to salinity changes was different as shown in (Figures 4.7 and 4.9). An incubation time of several hours in 0.03% salinity had no effect on the AOR and NOR (Figure 4.21 a, b), whereas, the incubation at higher NaCl concentration (5%) for a few hours resulted in decreasing of the NORs but not the AORs. On the other hand, an incubation time of about two weeks in 0.03% and also 5% salinity, as applied in FBRs C–I, revealed different results (Figures 4.7 and 4.9). These indicated that AOB and

NOB responded also differently to rates of salinity changes in the medium and a different ability to develop adaptation during incubation time seemed to prevail.

- *Batch assays with suspended nitrifiers (starting with 3.5% salt)*

AOR and NOR measurements during batch assays with suspended biomass were based on a quasi-linear slope at the beginning of the incubation (Figure 4.22a) in the same way as oxidation rates by biofilm nitrifiers were determined (Figure 4.6). A similar effect of salinity on NOR by suspended (Figure 4.23b) and biofilm nitrifiers (Figure 4.21b) was obtained, in which a few hours of incubation at low salinity did not influence NORs.

A severe effect was seen for AORs by suspended biomass (Figure 4.23a) compared to AORs by biofilm nitrifiers (Figure 4.21). In addition, a permanent effect on AOR was only found in assays with suspended biomass. The advantage of biofilm growth for treating saline wastewater, especially with fluctuating concentrations, was found in a previous study. However, in our case, other tests should be conducted to support the statement. The reason of that is the following

1. Magnetic bar mixing in assays with suspended biomass enables salt ions to penetrate deeper into flocs.
2. Based on our test (results not shown) mixing by different rates of aeration had almost no effect on AOB and NOB. This might be caused by an only thin biofilm in porous ceramic rings.
3. Although the original cultures for suspended and immobilized nitrifiers were the same, but both kinds of bacteria had a different culture history. The culture history can play a significant role in the observed effect of nitrification (Sharma and Ahler 1977).

To achieve the optimum salinity for conventional nitrification, the values of AOR:AOR $_{control}$ should be relative high with a minimal NAR (nitrite accumulation rate):AOR ratio and nitrite accumulation. Salinities of 3.5% and 1% were relatively good in accordance with the above criteria (Table 4.3). By maintaining a relatively high ratio of $AOR:AOR_{control}$ with a maximal NAR, the ideal condition for partial nitrification (nitritation) can be achived. These criteria could not be fullfilled in experiments with varying salinity.

- *Batch assays with suspended nitrifiers (starting with 0.03% salt)*

The activity of non halophilic nitrifiers, especially of nitrite oxidizers in a suspended growth system could not cope with salinity fluctuations (Figure 4.25). The inability of non halophilic nitrifiers was also found in the biofilm experiments (Figures 4.11 and 4.12).

At salinities of 0.5 and 1% (Table 4.4) $AOR:AOR_{control}$ ratios, NAR:AOR ratios and maxima of nitrite accumulation were higher than the values at a salinity of 0.03%. This indicated that partial nitrification by nonhalophilic nitrifiers could be achieved by increasing the salinity. The same indication was also found for application of nonhalophilic nitrifiers that formed a biofilm Figure 4.11, FBR K, phase V).

Table 5.3 The range of salinity fluctuation in our study resulting in 60–100% of initial AORs and NORs

Reactor	Initial salinity (%)	Time adaptation	Salinity range (%)		Recover ability (%)
			AOR	NOR	
FBR ceramic rings	3.5	About 14 days	0.03–5	1–5	>80
FBR Pelia	3.5	About 14 days	0.03–5	1–3.5	>80
FBR Pelia	0.03	About 14 days	0.03–5	0.03–1	>90
FBR ceramic rings	3.5	2–4 hours	0.03–5	0.03–5	>80
Suspension	3.5	2–4 hours	0.5–5	0.03–5	>70 (AOR) >95 (AOR)
Suspension	0.03	2–4 hours	0.03–2	0.03–1	Not tested

The rate measurement of nitrifying activity enables a direct comparison with other results. The range of salinity fluctuations in our study resulted in 60–100% of initial AORs and NORs and is summarized in Table 5.3, whereas the salt impacts on nitrification activity in selected reports are presented in Table 5.4.

Table 5.4 Reported results on the impact of salt on nitrification activity

Salt range	Nitrifiers seed	System used[a]	Salinity shift (%)	Reported impact in percentage compared to previous rate	Recovery	Literature.[b]
1.3–5	Activated sludge adapted to 2.6 and 3.3	FR	2.6 to 1.3 3.3 to 1.65 2.6 to 4 3.3 to 5	140 140 80 70 (Nitrification efficiency)	nm	1
0–3	Activated sludge non adapted	AS	0 to 0.5 to 1 to 3 0 to 7	70 to 51 to then 45 (AOR) 10 (AOR)	>90% in 5–10 days	2
0–3	Activated sludge adapted to 5%	AS	0.5 to 1to 3 0 to 7	79 to 70 (AOR) 20 (AOR)	>90% in 4–7 days	2
0–5	Nitrosomonas & Nitrobacter	AS	0 to 3 to 5 0 to 3 to 5	96 to then 80 (ammonia removal eff.) 90 to then 75 (AOR)	nm	3
0–6.6	Nitrifying activated sludge	SBR	At 6.6 0 to 1.65	Completely inhibited 45 (shock load, AOR) 61 (steady state, AOR) 58 (shock load, NOR) 95 (steady state, NOR)	40% in 1week - AOR 30% in 2 days NOR	4
0–1.02	Sewage	SBR	0 to 1.02 0 to 2	50 (ammonia removal eff.) 1 (AOR)	94% after 18th cycle	5
0–10.5	River mixture sludge-water	FBR	0 to 3.5 0 to 7	90 (AOR) 35 (NOR) 74 (AOR) 16 (NOR)	100% in 2 weeks - AOR & NOR	6
0–9	Marine mixture sludge-water	FBR	3.5 to 1 3.5 to 5	103 (AOR) 78 (NOR) 75 (AOR) 65 (NOR)	100% in 2 weeks - AOR >80% in 2 weeks- NOR	6

[a] FR = Fluidized reactor; FBR = Fixed bed reactor; AS = Activated sludge; SBR = Sequencing batch reactor.
[b] 1 = Dahl et al. 1997; 2 = Panswad and Anan 1999; 3 = Dincer and Kargi 2001; 4 = Moussa et al. 2006; 5 = Ye et al. 2009; 6 = this study

FBRs with porous ceramic rings or polyethylene/clay sinter lamellas show a similar trends of nitrification at 3.5 % salinity, indicating the selection of the same population of nitrifiers in both FBRs. Nitrification by halophilic or halotolerant bacteria (originating from marine water), but not by non-halophilic nitrifiers could cover a wider variation of salinity. However, the growth rate of non-halophilic nitrifiers in normal water is faster than that of halophilic nitrifiers in saline water. Nitrification plants handling wastewater with salinity fluctuations from 0–1%, therefore, could be inoculated with nitrifiers from fresh water or with non-halophilic nitrifiers.

5.4 Effect of temperature, ammonia, nitrite and nitrate on nitrification

- *Temperature*

The observed effects of temperature on AORs and NORs was the direct result of the influence of temperature on the reaction rates (Arrhenius relation) rather than a toxic effect of free ammonia (FA) and free nitrous acid (FNA). According to Equation 2.12 and 2.13, the concentration of FA and FNA was calculated to be 0.5 mg N L^{-1} and 8 x 10^{-4} mg N L^{-1} at 22.5°C, pH 8.0 for 10 mg L^{-1} NH_4^+-N or $N0_2^-$-N. According to Anthonisen et al. (1976) both concentrations should not have a significant inhibitory effect on AORs and NORs.

AOB and NOB have different growth rates at different temperatures, requiring correction factors for the overall nitrification rate. In this study, a decrease of AORs and NORs to about half of the initial rates at decreasing temperatures from 22.5 to 12.5°C (44% and 55%, respectively. Figure 4.28) is in good agreement with the Arrhenius equation.

It also can be said that the correction factor for abrupt decreasing temperature in this study was 1.059 for AOR and 1.047 for NOR. Whereas, Hwang and Oleszkiewics (2007) reported that for a slowly decreasing temperature the necessity of correction factors for nitrification ranged from

1.027 to 1.127 or for an abrupt temperature change ranged from 1.088 to 1.116. Correction factors of other reports are summarized in Table 5.5.

Table 5.5 Correction factors for decreasing nitrification temperatures

Nitrifiers used	Temperature range (°C)	Temperature correction factor		Literature
		AOR	NOR	
Biofilm	10–29	1.02	-	Fdz-Polanco et al. 1994
Biofilm	8–27*	1.043	-	Zhu and Chen 2002
Suspension	10–23	1.098	-	Salvetti et al. 2006
Suspension	10–20*	1.027 to 1.127	-	Hwang and
	10–20**	1.088 to 1.116		Oleszkiewicz 2007
Suspension	5–20	1.172	-	Guo et al. 2010
	20–35	1.062		
Biofilm	12.5–22.5°C**	1.059	1.047	This study
Biofilm	22.5–32.5°C	1.036	1.026	This study

* Gradual decreasing temperature, ** Abrupt decreasing temperature,

Temperature has a different influence on maximum growth rates of AOB and NOB. At temperatures of >15°C AOB have a higher growth rate than NOB (Bougard et al. 2006). At temperatures higher than 40°C AOB or AORs were less affected by the temperature than were NOB or NORs.

- *Ammonia, nitrite and nitrate*

The non-ionized ammonia (FA) and nitrous acid (FNA), rather than ammonia or nitrite ions were inhibitors of nitrification (Anthonisen et al. 1976). According to literature FA would cause inhibition of AOB in the range of 10 to 150 mg N L^{-1} and of NOB in the range of 0.1 to 270 mg N L^{-1} whereas the concentration range of FNA causing inhibition of AOB (0.002 to 0.5 mg N L^{-1}) and of NOB (0.22 to 2.8 mg N L^{-1}) is much lower (Table 5.6).

The different inhibitory concentrations listed in Table 5.6 seem to be the result of different experimental conditions such as pH value, temperature, salinity, the cultivation method (batch assays or continuous reactors) and of the used suspended or immobilized pure or mixed cultures as listed in the legend of the table. Another possible explanation for different and respectively higher inhibitory concentrations is the adaption phenomenon which was reported by Villaverde et al. (2000) and Qiao et al. (2010). Acclimatization to higher loads of ammonia or nitrite containing wastewater could be the reason for higher initial inhibitory concentrations of FA and FNA.

Table 5.6 Inhibition initiating concentrations of FA and FNA for AOB and NOB

	FA mg N L^{-1}	FNA mg N L^{-1}	Literature
AOB	10–150[a]		Anthonisen et al. 1976
	> 10[b]		Mosquera-Corral et al. 2005
	78[c*]		Kim et al. 2006
	54[d*]		this study
		0.1[e]	Vadivelu et al. 2006
		0.5[f]	Qiao et al. 2010
		0.0021[d**]–0.02[d***]	this study
NOB	0.1–10[a]		Anthonisen et al. 1976
	> 1[b]		Mosquera-Corral et al. 2005
	0.7[c**]		Kim et al. 2006
	270[d****]		this study
		0.22–2.8[a]	Anthonisen et al. 1976
		>> 0.79[d*****]	this study

[a] : Initiating concentration, 10–23°C, pH: adjusted purposefully, low salinity, batch and continuous reactor, mixed culture
[b] : Initiating concentration, 35°C, pH 6.8–7.5, low salinity, continuous stirred tank reactor, mixed culture
[c*] : Completed inhibition, 25°C, pH 8.2, low salinity, batch reactor, mixed culture
[c**] : IC_{50}, 25°C, pH 8.2, low salinity, continuous biofilm airlift reactor, mixed culture
[d*] : IC_{50}; [d**] : IC_{50}; [d***] : 100% inhibition; initiating concentration; [d****] : IC_{50} ; [d*****] : no inhibition, 22.5°C, pH 8, 3.5% salinity, batch reactor, mixed culture
[e] : Initiating concentration, 30°C, pH 7.1, low salinity, batch reactor, mixed culture
[f] : Initiating concentration, 21°C, pH 6.5, low salinity, continuous swim bed reactor, mixed culture

The mechanisms responsible for the FA and FNA inhibition on the respiration of *Nitrobacter* sp. are not clear. It may be due to a direct inhibitory effect of FA and FNA on enzymes involved in the electron transport or proton translocation (Vadivele et al. 2006). The measurement of activity of mixed cultures such as ammonia and nitrite oxidizers is a difficult task. The addition of allylthiourea and sodium azide as specific inhibitors allows the separate measurement of effects of ammonia on nitritation and nitratation (Ginested et al. 1998). The substances were also used by Rongsayamanont et al. (2010) and Sanches et al. (2001). In our study, only allythiourea functioned as specific inhibitor (Figure 4.20a).

Furthermore, it was suggested that the inhibitory effect on energy generation (catabolic processes) and growth processes (anabolic processes) should be investigated separately (Vadivelu et al. 2006). Simply comparing the overall substrate utilization rates, as conducted in our study, might have seriously under-estimated the inhibitory effects, certainly in relation to growth inhibition.

5.5 Partial nitrification

Recently, many methods and approaches have been reported to achieve partial nitrification. Regulations of the reactor such as temperature, pH, dissolved oxygen concentration, sludge retention time, aeration pattern, and substrate concentration have been suggested.

The biofilm formed in the inner tube wall of the external recirculation tube in FBR A and FBR B, in fact, was dominated by ammonia oxidizing bacteria (Figure 4.14). The flow velocity (0.028 m s^{-1}) of the medium through the tube resulted in high turbulence, allowing AOB but not NOB to attach at the wall of the rubber tube. Another explanation of AOB but not NOB growth at the inner wall of rubber tubes is that air could diffuse through the tube and supply oxygen for the bacteria, as experimentally shown by Hsieh et al. (2002). This might provide a condition where only AOB could growth in the silicon tube.

During biomass selection (Figure 4.1a) and acclimatization in FBRs A and B (Figure 4.2), the activity and growth of AOB seemed to be faster than that of NOB. The different growth rates of both bacteria types in activated sludge enables a washout of NOB from reactors by regulating sludge retention times (Hellinga et al. 1998). To date, no method has been reported to washout only NOB from attached growth nitrifying reactors.

Increasing of salinity could suppress more NOR than AOR (Figures 4.9 and 4.12). Indeed, ammonia, nitrite but no nitrate was found in the effluent from FBRs inoculated with non-halophilic nitrifiers, when salinity was increased from 0.03 to 9%, indicating that NOR was completely suppressed by such high salinity (Figure 4.11, FBR K, phase V). Similar results were also obtained from phase VI of FBR J (Figure 4.11). Salt inhibition at 5 g L^{-1} has been used by Ye et al. (2009) to achieve the partial nitrification.

Conditions of partial nitrification are required for the anammox process, and the NO_2^--N/NH_4^+-N ratio produced by partial nitritation should be around 1.0. When the ALR in FBR O was increased from 85 to 125 mg N L^{-1} d^{-1} the total nitrogen from ammonia (about 62.5 mg L^{-1} NH_4^+-N) and nitrite (about 62.5 L^{-1} NO_2^--N) in effluent was the same as the total nitrogen in the influent (125 mg L^{-1} NH_4^+-N) indicating that NOB were suppressed and this led to a required condition for the anammox process to be established. Liang et al. (2011) reported that 50% partial nitrification could be achieved by stepwise increases of influent ammonia.

During the measurement of AORs in batch assays, an increase of nitrite accumulation was found at increasing temperature, indicating that NOB were more suppressed than AOB (data not shown). Hellinga et al. (1998) also reported that above 30°C the growth rate of AOB was faster than that of NOB.

Chapter 6

CONCLUSION

- *Microbial analysis*

Seawater sample from Hafen Büsum contained nitrifiers having activity in saline environment. The sample was characterized by a conductivity of 31.9 mS cm^{-1}, an alkalinity of 120 mg L^{-1} as $CaCO_3$, an ammonia concentration of 0 mg NH_4^+-N L^{-1}, a nitrite concentration of 0.11 mg NO_2^--N L^{-1} and a nitrate concentration of 2.3 mg NO_3^--N L^{-1}.

During "activation" of nitrification nitrifiers in the samples needed a lag phase of about 10 days for ammonia oxidation and a longer lag phase for nitrite oxidation to proceed. The sample that contained relative little mud had an ammonia oxidation rate (AOR) of 11.5 mg N day^{-1}. This sample, then, was selected as an inoculation for the polyethylene/clay fixed bed reactor (FBR A) and for the porous ceramic ring fixed bed reactor (FBR B).

Molecular biological analyses after applying the Polymerase chain reaction (PCR) to sludge samples from FBR B revealed that the biomass in the reactor was dominated by *Nitrosomonas aestuarii* and *Nitrosomonas* Nm143 as AOB and *Nitrospira marina* as NOB. Such DNA sequences were usually found in coastal and marine water samples or sediments.

- *Feasibility of fixed bed reactors for nitrification of saline wastewater*

More than 300 days FBR A and FBR B, which were fed with synthetic saline wastewater that contained 3.5% NaCl, run continuously at different conditions concerning the hydraulic rate time (HRT), ammonia loading rate (ALR) and external water recirculation.

The support materials in the FBRs (polyethylene/clay and porous ceramic rings) have an important role during the early stage of biofilm formation. The roughness of surface due to crevices and pores of the supporting material facilitate fast bacteria colonization.

Porous ceramic rings applied in FBR B have a specific surface area of 934 m^2 m^{-3}, twice as high a specific surface area than polyethylene/clay in FBR A. Better nitrification performance was observed in FBR B. Incomplete ammonia oxidation only occurred when pH titration or aeration failed.

In almost all measurements, the NOR was always higher than the AOR in both reactors. The maximum AORs and NORs before installation of the liquid recirculation were 6 and 7 mg N L^{-1} h^{-1}, respectively, in FBR A and 5 and 5.9 mg N L^{-1} h^{-1} in FBR B. Then, maximum area-related AOR or NOR, which was calculated by dividing the maximum N removal by the specific surface area of the support material, was 312 and 386 mg N m^{-2} day^{-1} for FBR A and 199 and 236 mg N L^{-1} day^{-1} for FBR B, respectively.

The NOR, especially in FBR B, was relatively stable, whereas the AOR showed a higher fluctuation with time.

A better homogenization of medium and improved mass transfer of oxygen and ammonia achieved by applying external recirculation resulted in a short term increase of the AOR (>2 fold) and the NOR (1.5 fold) in FBR A, but had no influence on FBR B. On the other side, the application of external recirculation initiated a biofim detachment in both reactors.

Biofilm on supporting material in FBR A and B was distributed uneven. Dense but not compact biofilm was visible inside the rings and on top of the rings in FBR B. In other surface areas, where shear stress was high, the biofilm was invisible, but nitrification activity was present.

The average decay rate of the nitrifiers under aerobic conditions of both reactors at 22.5°C was relatively low, being 0.03 d^{-1} for AOB and 0.045 d^{-1} for NOB. At the same temperature the average decay rate of NOB under anaerobic conditions was 0.048 d^{-1}. If a complete interruption of reactor operation in a reactors for revision was necessary for one week, for example, the activity of biofilm nitrifiers on porous ceramic rings or polyethylene/clay sinter lamellas was still approx 80% for AOB and 65% for NOB of initial activity.

- *Effect of salinity changes on nitrification*

At a fixed HRT of 1 day, during the salinity changes in fixed bed reactors C-N, the salinity in the reactors reached the new concentration after 1-3 days.

The gradual change of salinities could be tolerated. If the salinity was changed from 3.5 % to not lower than 0.5 % or not higher than 5 % the reactor with starting with 3.5% salt maintained at least 60 % of its original performance. Whereas the FBRs inoculated with fresh water sludge could be operated in a salinity range of 0.03–1% with a 40% decrease of initial AORs and NORs.

The salinity fluctuation influenced more NOBs than AOBs during continuous reactor operation. It was observed that accumulation of nitrite but not of ammonia at the start of these phase changes occured. Besides, during batch incubation NORs were more sensitive to the salinity fluctuation than AORs.

FBRs inoculated with seawater sludge can operate a wider range of salinity. The AOR remained constant after the salinity was decreased from 3.5% to 2, 1 and 0.5%. After the salt concentration was brought to 0.034%, the AOR was approx. 70% of initial rate. Whereas, AOR of FBRs inoculated with freshwater sludge and starting with 0.03% salt was not influenced, when salinity was increased from 0.03% to 0.5, 1 and 2%. The AOR in the FBRs was still approx. 90% of initial rate after the salt concentration was increased to 3.5%.

On the other side, a decrease of the NORs in FBRs with 3.5% salt starting concentration occurred already when the salinity was decreasing from 3.5 % to 2 %. The NORs were 62, 76 and 60% of the initial rates after salinity was decreased from 3.5% to 2, 1 and 0.5%, respectively. The salinity effect on NORs in FBRs with starting concentrations of 0.03% salt was already found after salinity increased to 1%. The NORs were 85, 52 and 36% of the initial rates after salinity was increased to 1, 2 and 3.5%, respectively.

The transition from 3.5 % NaCl towards a hypotonic environment (lower salt concentration than before) was less destructive for nitrification than a transition towards a hypertonic environment (higher salt concentration than before).

The average AORs and NORs after the salinity was increased from 3.5% to 5, 7 and 9% were approx. 68.5, 52.5 and 36% (for AOR) and 87.5, 56 and 46.5% for (NOR) of initial rates, respectively. AORs and NORs in FBRs inoculated with freshwater sludge were almost zero, when the salinity was increased above 9%.

Returning to the 3.5 % and 0.03% original salt concentration revealed a long term restoration of the full activity of AOB and NOB.

- *Effect of ammonia, nitrite and nitrate on nitrification*

Free ammonia (FA) and free nitrous acid (FNA), rather than ammonia or nitrite ions were inhibitors of nitrification. The ammonia and nitrite concentration should be converted to non-ionized ammonia FA and FNA.

FNA was more toxic for both, AOB and NOB than FA.

AORs were more affected by FA than NORs, so that concentration of 0.25-0.5 g N L^{-1} (calculated FA of 13.5 - 27.2 mg N L^{-1}) was sufficient to initiate inhibition, whereas NOR inhibition was initiated at 3.5 g N L^{-1} (calculated FA of 190 mg N L^{-1}). Ammonia IC_{50} for AOR and NOR was about 1 g N L^{-1} (calculated FA of 54 mg N L^{-1}) and 5 g N L^{-1} (calculated FA of 270 mg N L^{-1}).

AORs were also more sensitive to FNA than NORs. Nitrite inhibition concentrations of 50% (IC_{50}) and 100% inhibition for AORs were about 0.027 g N L^{-1} (calculated FNA of 0.0021 mg N L^{-1} and 0.25 g N L^{-1}, respectively. No inhibition on NOR was observed when nitrite, even up to 10 g N L^{-1} (calculated FNA of 0.79 mg N L^{-1}) was applied to the medium. No inhibition on AOR and NOR by nitrate up 10 g N L^{-1} was found.

The activity recovery of AOB and NOB was possible indicating the temporary effect of ammonia and nitrite inhibition.

- *Effect of temperature on nitrification*

A direct rather than an indirect effect of temperature was observed on AOR and NOR. The AOR was slightly more sensitive than the NOR for temperature fluctuation.

A temperature decrease of 10 °C (from 22.5 to 12.5 °C) in the presence of 3.5% NaCl revealed a decrease of AOR and NOR by 44% and 55% (about half of the initial rate), respectively, whereas a temperature increase of 10 °C (from 22.5 to 32.5 °C) resulted in an increase of the AOR and NOR to 136% and 126%, which was not a doubling of nitritation and nitratation rates.

A further decrease of the temperature to 6 °C, simulating nitrification temperatures during the winter season prevented ammonia oxidation completely and reduced the NOR by more than 90 %. A further increase of temperature up to 40°C did not affect the rates for both processes and a temperature of 50°C resulted in no activity of both bacteria. A full recovery of the AOR and NOR was obtained after the ‘‘cold shock’’ by increasing the temperature from 6 to 22.5°C, however no recovery of AOR or NOR was obtained after the ‘‘heat shock’’ when the temperature was reduced from 50°C to 22.5°C.

REFERENCES

Ahn, Y.H., 2006. Sustainable nitrogen elimination biotechnologies: A review. Process Biochem. 41, 1709-1721.

Anthonisen, A.C., Loehr, R.C., Prakasam, T.B.S., Srinath, E.G., 1976. Inhibition of nitrification by ammonia and nitrous acid. J. Water Pollut. Control Fed.. 48(5), 853–852.

Antileo, C., Aspe, E., Urrutia, H., Zaror, C., Roeckel, M., 2002. Nitrifying biomass acclimation to high ammonia concentration. J. Environ. Eng. 128, 367–375.

American Public Health Association (APHA), 1985. Standard Methods for the Examination of Water and Wastewater. 15th Edition. Washington, D.C.

Aspe, E., Marti, M.C., Roeckel, M., 1997. Anaerobic treatment of fishery wastewater using a marine sediment inoculum. Water Res. 31(9), 2147–2160.

Beccari, M., Di-Pinto, A.C., Ramadori, R., Tomei, M.C., 1992. Effects of dissolved oxygen and diffusion resistances on nitrification kinetics. Water Res. 26(8), 1099–1104.

Behrendt, H., Kornmilch, M., Opitz, D., Schmoll, O., Scholz, G., 2002. Estimation of the nutrient inputs into river systems – experiences from German rivers. Reg. Environ. Change. 3, 107–117.

Biesterfeld, S., Farmer, G., Russell, P., Figueroa, L., 2003. Effect of alkalinity type and concentration on nitrifying biofilm activity. Water Environ. Res. 75(3), 196–204.

Biggs, C.A., Lant, P.A., 2000. Activated sludge flocculation: on-line determination of floc size and the effect of shear. Water Res. 34(9), 2542–2550.

Boller, M., Gujer, W., Tschui, M., 1994. Parameters affecting nitrifying biofilm reactors. Water Sci. Technol. 29(10–11), 1–11.

Bougard, D., Bernet, N., Chèneby, D., Delgenès, J.-P., 2006. Nitrification of a high-strength wastewater in an inverse turbulent bed reactor: Effect of temperature on nitrite accumulation. Process Biochem. 41, 106–113.

Campos, J.L., Mosquera-Collal, A., Sanchez, M., Méndez, R., Lema, J.M., 2002. Nitrification in saline wastewater with high ammonia concentration in an activated sludge unit. Water Res. 36, 2555–2560.

Carrera, J., Baeza, J.A., Vicent, T., Lafuent, J., 2003. Biological nitrogen removal of high-strength ammonium industrial wastewater with two-sludge system. Water Res. 37, 4211–4221.

Catalan-Sakairi, A.B.M., Wang, P.I.C., Matsumura, M., 1997. Nitrification performance of marine nitrifiers immobilized in polyester- and macro-porous cellulose carriers. J. Ferment. Bioeng. 84(6), 563–571.

Chae, K.J., Rameshwara, T., Jang, A., Kim, S.H., Kim, I.S. 2008. Analysis of the nitrifying bacterial community in biocube sponge media using fluorescent in situ hybridization (fish) and microelectrodes. J. Environ. Manage. 88, 1426–1435.

Characklis, W.G., Trulear, M.G., Briyers, J.D., Zelver, N., 1982. Dynamics of biofilm processes: methods. Water Res. 16, 1207-1216.

Chen, S., Ling, J., Blancheton, J.P., 2006. Nitrification kinetics of biofilm as affected by water quality factors. Aquac. Eng. 34, 179–197.

Dahl, C., Sund, C., Kristensen, G.H., Vredenbregt, L., 1997. Combined biological nitrification and denitrification of high-salinity wastewater. Water Sci. Technol. 36(2–3), 345–352.

Daims, H., Nielsen, JL., Nielsen, PH., Schleifer, KH., Wagner, M., 2001. In situ characterization of Nitrospira-like nitrite-oxidizing bacteria active in wastewater treatment plants. Appl. Environ Microbiol. 67, 5273–5284.

Dalton, H.M., Goodman, A.E., Marshall, K.C., 1996. Diversity in surface colonization behaviour in marine bacteria. J. Ind. Microbiol. 17, 228–234.

De Beer, D., Stoodley, P., Lewandowski, Z., 1996. Liquid flow and mass transport in heterogeneous biofilm. Water Res. 30(11), 2761–2765.

DEV 1983. Deutsche Einheitsverfahren zur Wasser-, Abwasser- und Schlammuntersuchung. Chemie, Weinheim.

Dincer, A.R., Kargi, F., 2000. Kinetics of sequential nitrification and denitrification processes. Enzyme Microb. Technol. 27, 37–42.

Dincer, A.R., Kargi, F., 2001. Salt inhibition kinetics in nitrification of synthetic saline wastewater. Enzyme Microb. Technol. 28, 661–665.

Dincer, AR., Kargi, F., 2001. Performance of rotating biological disc system treating saline wastewater. Process Biochem. 36, 901–906.

Elenter, D., Milferstedt, K., Zhang, W., Hausner, M., Morgenroth, E., 2007. Influence of detachment on substrate removal and microbial ecology in a heterotrophic/ autotrophic biofilm. Water Res. 41, 4657–4671.

Fdz-Polanco, Villaverde, S., Garcia, P.A., 1994. Temperature effect on nitrifying bacteria activity in biofilters: activation and free ammonia inhibition. Water Sci. Technol. 30(11), 121–130.

Fitch, MW., Pearson, N., Richard, G., Burken, JG., 1998. Biological fixed film system. Water Environ. Res. 70, 495–518.

Fontenot, Q., Bonvillain, C., Kilgen, M., Boopathy, R., 2007. Effects of temperature, salinity, and carbon: nitrogen ratio on sequencing batch reactor treating shrimp aquaculture wastewater. Bioresour. Technol. 98, 1700–1703.

Fox, P., Suidan, MT., Bandy, JT., 1990. A comparison of media types in acetate fed expanded-bed anaerobic reactor. Water Res. 24, 827–835.

Francis, CA., Roberts, KJ., Beman, JM., Santoro, AE., Oakley, BB., 2005. Ubiquity and diversity of ammonia-oxidizing archaea in water columns and sediments of the ocean. Proc Natl Acad. Sci. USA 102, 14683–14688.

Furukawa, K., Ike, A., Ryu, S.L., Fujita, M., 1993. Nitrification of NH_4^+-N polluted sea water by immobilized acclimatedmarine nitrifying sludge (AMNS). J. Ferment. Bioeng. 76(6), 515–520.

Galinski, E.A., Trüper, H.G., 1994. Microbial behaviour in salt-stressed ecosystems. FEMS Microbiol. Rev. 15(2–3), 95–108.

Gharsallah, N., Khannous, L., Souissi, N., Nasri, M., 2002. Biological treatment of saline wastewaters from marine-products processing factories by a fixed-bed reactor. J. Chem. Technol. Biotechnol. 77, 865–870.

Ginestet, P., Audic, J.M., Urbain, V., Block, J.C., 1998. Estimation of nitrifying bacterial activities by measuring oxygen uptake in the presence of the metabolic inhibitors allylthiourea and azide. Appl. Environ. Microbiol. 64(6), 2266–2268.

Gjaltema, A., van der Marel, N., van Loosdrecht, MCM., Heijnen, J.J., 1997. Adhesion and biofilm development on suspended carriers in airlift reactor: hydrodynamic conditions versus surface characteristics. Biotechnol. Bioeng. 55, 880–889.

Graaf, A. A. van de., Mulder, A., Bruijn, P., de, Jetten, M. S. M., Robertson, L. A., and Kuenen, J.G., (1995). Anaerobic ammonium oxidation is a biologically mediated process. Appl. Environ. Microbiol. 61, 1246–1250.

Grady, J.C.P.L., Lim, H.C., 1980. Biological wastewater treatment: Theory and application. Maracell Dekker, Inc., New York.

Guo, J., Peng, Y., Huang, H., Wang, S., Ge, S., Zhang, J., Wang, Z,. 2010. Short- and long-term effects of temperature on partial nitrification in a sequencing batch reactor treating domestic wastewater. J. Hazard. Mater. 179, 471–479.

Hamdi, M., 1995. Biofilm thickness effect on the diffusion limitation in the bioprocess reaction: biofloc critical diameter significance. Bioprocess Eng. 12, 193–1997.

HELCOM, 2009. Eutrophication in the Baltic Sea–An integrated thematic assessment of the effects of nutrient enrichment and eutrophication in the Baltic Sea region: Executive Summary. Balt. Sea Environ. Proc. No. 115A.

Hellinga, C., Schellen, A.A.J.C., Mulder, J.W., van Loosdrecht, M.C.M., Heijnen, J.J., 1998. The sharon process: An innovative method for nitrogen removal from ammonium-rich waste water. Water Sci. Technol. 37(9), 135–142.

Hellinga, C., Schellen, A.A.J.C., Mulder, J.W., van Loosdrecht, M.C.M., 1998. The Sharon process: an innovative method for nitrogen removal from ammonium-rich wastewater. Water Sci. Technol. 37, 135–142.

Hsieh, Y.L., Tseng, S.K., Chang, Y.J., 2002. Nitrification using polyvinyl alcohol-immobilized nitrifying biofilm on an O_2-enriching membrane. Biotechnol. Lett. 24, 315–319.

Huilinir, M., Cristina Marti, M., Aspe, E., Roecke, M., 2008. Organic and nitrogenous matter effects on the denitrification of saline protein rich effluents. Environ. Technol. 29, 881–890.

Hwang, J.H., Oleszkiewicz, J.A.,2007. Effect of cold-temperature shock on nitrification. Water Environ. Res. 79, 964–968.

Imhoff , JF., Thiemann, B., 1991. Influence of salt concentration and temperature on the fatty acid composition of Ectothiorhodospira and other halophilic phototrophic purple bacteria. Arch. Microbiol. 156, 370–375.

Intrasungkha, N., Keller, J., Blackall, L.L., 1999. Biological nutrient removal efficiency in treatment of saline wastewater. Water Sci. Technol. 39(6), 183–190.

Jorand, M.F., Zartarian, F., Thomas, M.F., Block, J.C., Bottero, M.J.Y., Villemin, G., Urbain, V., Manem, J., 1995. Chemical and structural (2D) linkage between bacteria within activated sludge flocs. Water Res. 297, 1639–1647.

Kawasaki, S., Gromiec, M.J., Hanaki, K., Matsumoto, J., 1996. Effect of seawater on nitrification by attached biofilm. Ecol. Modell. 32, 183–190.

Keiding, K., Nielsen, P.H., 1997. Desorption of organic macromolecule from activated sludge: effect of ionic composition. Water Res. 31(7), 1665–1672.

Kim, D.J., Lee, D.-I., Keller, J., 2006. Effect of temperature and free ammonia on nitrification and nitrite accumulation in landfill leachate and analysis of its nitrifying bacterial community by FISH. Bioresour. Technol. 97, 459–468.

Kim, J.H., Guo, X., Park, H.S., 2008. Comparison study of the effects of temperature and free ammonia concentration on nitrification and nitrite accumulation. Process Biochem. 43, 154–160.

Koops, H.P., Böttcher, B., Möller, U.C., Pommerening-Röser, A., Stehr, G., (1991). Classification of eight new species of ammonia-oxidizing bacteria: Nitrosomonas communis sp. nov., Nitrosomonas ureae sp. nov., Nitrosomonas aestuarii sp. nov., Nitrosomonas marina sp. nov., Nitrosomonas nitrosa sp. nov., Nitrosomonas eutropha sp. nov.,

Nitrosomonas oligotropha sp. nov., and Nitrosomonas halophila sp. nov., J. Gen. Microbiol. 137, 1689–1699.

Krüner, G., Rosentha, H., 1983. Efficiency of nitrification in trickling filters using different substrates. Aquacult Eng. 2, 49–67.

Lazavora, V., Manem, J., 1995. Biofilm characterization and activity analysis in water and wastewater treatment. Water Res. 29(10), 2227–2245.

Lefebvre, O., Vasudevan, N., Torrijos, M., Thanasekaran, K., Moletta, R., 2005. Halophilic biological treatment of tannery soak liquor in a sequencing batch reactor. Water Res. 39, 1471–1480.

Liang, Z., Han, Z., Yang, S., Liang, X., Du, P., Liu, G., Yang, Y., 2011. A control strategy of partial nitritation in a fixed bed biofilm reactor. Bioresour. Technol. 102, 710–715.

Liao, B.Q., Allen, D.G., Leppard, G.G., Droppo, I.G., Liss, S.N., 2002. Interparticle interactions affecting the stability of sludge flocs. J. Colloid Interface Sci. 249(2), 372–380.

Liu, S., Yang, F., Gong, Z., Su, Z., 2008. Assessment of the positive effect of salinity on the nitrogen removal performance and microbial composition during the start-up of CANON process. Appl. Microbiol. Biotechnol. 80, 339–348.

Mariangel, L., Aspe, E., Cristina Marti, M., Roeckel, M., 2008. The effect of sodium chloride on the denitrification of saline fishery wastewater. Environ. Technol. 29, 871–879.

Metcalf & Eddy, Inc., 2003. Wastewater Engineering: Treatment and reuse 4th ed. New York: McGraw-Hill.

Mobarry, B.K., Wagner, M., Urbain, V., Rittmann, B.E., Stahl, D.A., 1996. Phylogenetic probes for analyzing abundance and spatial organization of nitrifying bacteria. Appl. Environ. Microbiol. 62, 2156–2162.

Morgan, J.W., Forster, C.F., Evison, L., 1990. A comparative study of the nature of biopolymers extracted from an aerobic and activated sludges. Water Res. 24(6), 743–750.

Mosquera-Corral, A., Gonzalez, F., Campos, J.L., Mendez, R., 2005. Partial nitrification in a SHARON reactor in the presence of salts and organic carbon compounds. Process Biochem. 40, 3109–3118.

Moussa, M.S., Sumanasekera, D.U., Ibrahim, S.H., Lubberding, H.J., Hooijmans, C.M., Gijzen, H.J., Van Loosdrecht, M.C.M., 2006. Long term effects of salt on the activity, population structure and floc characteristics in enriched bacterial cultures for nitrifiers. Water Res. 40, 1377–1388.

Nijhof, M., Bovendeur, J., 1990. Fixed film nitrification characteristic in sea water recirculation fish culture systems. Aquac. 87, 133–134.

Nogueira, R., Lazarova, V., Manem, J., Melo, L.F., 1998. Influence of dissolved oxygen on the nitrification kinetics in a circulating bed biofilm reactor. Bioprocess. Eng. 19, 441–449.

Ohashi, A., Viraj de Silva, D.G, Mobarry, B., Manem, J.A., Stahl, D.A., Rittman, B.E., 1995. Influence of substate C/N ration on the structure of multi-species biofilms consisting of nitrifiers and heterotrophs. Water Sci. Technol. 32(8), 75–84.

Okabe, S., Satoh, H., Watanabe, Y., 1999. In situ analysis of nitrifying biofilms as determined by in situ hybridization and the use of microelectrodes. Appl. Environ. Microbiol. 65, 3182–3191.

Ollivier, B., Caumette, P., Garcia, J.L., Mah, R.A., 1994. Anaerobic bacteria from hypersaline environments. Microbiol. Rev. 58, 27–38.

Oren, A., 1999. Bioenergetic Aspects of Halophilism. Microbiol. Mol. Biol. Rev. June, 334–348.

OSPAR (Oslo and Paris) Commission, 2008. Second integrated report on the eutrophication status of the OSPAR Maritime Area.

Palenzuela-Rollon, A., Zeeman, G., Lubberding, H.J., Lettinga, G., Alaerts, G.J., 2002. Treatment of fish processing wastewater in a one- or two-step upflow anaerobic sludge blanket (UASB) reactor. Water Sci. Technol. 45(10), 207–212.

Panswad, T., Anan, C., 1999. Specific oxygen, ammonia, and nitrate uptake rates of a biological nutrient removal process treating elevated salinity wastewater. Bioresour. Technol. 70, 237–243.

Peng, Y.Z., Zhu, G.B., 2006. Biological nitrogen removal with nitrification and denitrification via nitrite pathway. Appl. Microbiol. Biotechnol. 73, 15–26.

Purkhold, U., Wagner, M., Timmermann, G., Pommerening-Röser, A., Koops, HP., 2003. 16S rRNA and amoA-based phylogeny of 12 novel betaproteo bacterial ammonia-oxidizing isolates: extension of the data set and proposal of a new lineage within the nitrosomonads. Int. J. Syst. Evol. Microbiol. 53, 1485–1494.

Qiao, S., Matsumoto, N., Shinohara, T., Nishiyama, T., Fufii, T., Bhatti, Z., Furukawa, K., 2010. High-rate partial nitrification performance of high ammonium containing wastewater under low temperatures. Bioresour. Technol. 101, 111–117.

Rejish Kumar, V.J., Achuthan, C., Manju, N.J., Philip, R., Bright Singh, I.S., 2009. Mass production of nitrifying bacterial consortia for the rapid establishment of nitrification in saline recirculating aquaculture systems. World J. Microbiol. Biotechnol. 25, 407–414.

Rongsayamanont, C., Limpiyakorn, T., Law, B., Khan, E., 2001. Relationship between respirometric activity and community of entrapped nitrifying bacteria: Implications for partial nitrification. Enzyme Microb. Technol. 46(3–4), 229–236.

Rosa, M.F., Furtado, A.A.L., Albuquerque, R.T., Leite, S.G.F., Medronho, R.A., 1998. Biofilm development and ammonia removal in the nitrification of a saline wastewater. Bioresour. Technol. 65, 135–138.

Salem, S., Moussa, M.S., Van Loosdrecht, M.C.M., 2005. Determination of the decay rate of nitrifying bacteria. Biotechnol. Bioeng. 94(2), 252–262.

Sanches, O., Marti, M.C, Aspe, E., Roeckel, M. 2001. Nitrification rates in a saline medium at different dissolved oxygen concentrations. Biotechnol. Lett. 23, 1597–1602.

Sanchez, O., Aspe, E., Marti, M.C., Roeckel, M., 2005. Rate of ammonia oxidation in a synthetic saline wastewater by a nitrifying mixed-culture. J. Chem. Technol. Biotechnol. 80, 1261–1267.

Seo, J.K., Jung, I.H., Kim, M.R., Kim, B.J., Nam, S.W., Kim, S.K., 2001. Nitrification performance of nitrifiers immobilized in Pva (Polyvinyl Alcohol) for a marine recirculating aquarium system. Aquac. Eng. 24, 181–194.

Sharma, B., Ahler, R.C., 1977. Nitrification and nitrogen removal. Water Res. 11, 897–925.

Sheng, G.P., Yu, H.Q., Li, X.Y., 2006. Stability of sludge flocs under shear conditions: roles of extracellular polymeric substances (EPS). Biotechnol. Bioeng. 93(6), 1095–1102.

Siegrist, H., Gujer, W., 1987. Demonstration of mass transfer and pH effects in a nitrifying biofilm. Water Res. 21(12), 1481–1487.

Sliekers, A.O., Third, K.A., Abma, W., Kuenen, J.G., Jetten M.S.M., 2003. CANON and anammox in a gas-lift reactor. FEMS Microbiol. Lett. 218, 339–44.

Stenstrom, M.K., Poduska, R.A., 1980. The effect of dissolved oxygen concentration on nitrification. Water Res. 14, 643–649.

Stoodley, P., Saure, K., Davies, D.G., Costerton, J.W. 2002. Biofilms as complex differentiated communities. Annu. Rev. Microbiol. 56, 187–209.

Sudarno, U., Bathe, S., Winter, J., Gallert, C., 2010. Nitrification in fixed-bed reactors treating saline wastewater. Appl. Microbiol. Biotechnol. 85, 2017–2030.

Tang, S.L., Yue, D.P.T., Ku, D.C.C., 2007. Engineering and costs of dual water supply systems. IWA Publishing. UK.

Telgmann, U., Horn, H., Morgenroth, E., 2004. Influence of growth history on sloughing and erosion from biofilms. Water Res. 38, 3671–3684.

Tenney, M. W., and Stumm W. J., 1965. Chemical flocculation of microorganisms in biological waste treatment. J. Wat. Pollut. Control Fed. 39, 1370–1388.

Tijhuis, L., van Loosdrecht, M.C.M., Heijnen, J.J., 1994. Formation and growth of heterotrophic aerobic biofilms on small suspended particles in airlift reactors. Biotechnol. Bioeng. 44, 595–608.

UNEP, 2006. Challenges to International Waters – Regional Assessments in a Global Perspective. United Nations Environment Programme, Nairobi, Kenya.

Uygur A., 2006. Specific nutrient removal rates in saline wastewater treatment using sequencing batch reactor. Process Biochem. 4, 61–66.

Uygur, A., Kargi, F., 2004. Salt inhibition on biological nutrient removal from saline wastewater in a sequencing batch reactor. Enzyme Microb. Tech. 34, 313–318.

Vadivelu, V.M., Keller, J., Yuan, Z., 2006. Effect of free ammonia and free nitrous acid concentration on the anabolic and catabolic processes of an enriched Nitrosomonas culture. Biotechnol. Bioeng. 95(5), 830–839.

van Benthum,W.A.J., van Loosdrecht, M.D.M., Heijnen, J.J., 1997. Control of heterotrophic layer formation on nitrifying biofilms in a biofilm airlift suspension reactor. Biotechnol. Bioeng. 53, 397–405.

Verran, J., Lees, G., Shakespeare, A.P., 1991. The effect of surface roughness on the adhesion of candida albicans to acrylic. Biofouling. 3, 183–192.

Vieira, M.J., Melo, L.F., 1999. Intrinsic kinetics of biofilms formed under turbulent flow and low substrate concentration. Bioprocess Eng. 20(4), 369–375.

Villaverde, S., Fdz-Polanco, F., Garcia, P.A., 2000. Nitrifying biofilms acclimation to free ammonia in submerged biofilters: Start-up influence. Water Res. 34, 602–610.

Villaverde, S., Garcia-encina, P.A., Fdz-Polanco, F., 1997. Influence of pH over nitrifying biofilm activity in submerged biofilters. Water Res. 31, 1180–1186.

Vredenbregt, L.H.J., Nielsen, K., Potma, A.A., Kristensen, G.H., 1997. Fluid bed biological nitrification and denitrifacation in high salinity wastewater. Water Sci. Technol. 36(1), 93–100.

Wagner, M., Loy, A., 2002. Bacterial community composition and function in sewage treatment systems. Curr. Opinion Biotechnol.13(3), 218–227.

Wagner, M., Rath, G., Amann, R., Koops, H.P., Schleifer, K.H., 1995.In situ identification of ammonia-oxidizing bacteria. Syst. Appl. Microbiol. 18(2), 251–264.

Wang, X., Zhang, H., Yang, F., Wang, Y., Gao, M., 2008. Long-term storage and subsequent reactivation of aerobic granules. Bioresour. Technol. 99, 8304–8309.

Wäsche, S., Horn, H., Hempel, D.C., 2002. Influence of growth conditions on biofilm development and mass transfer at the bulk/biofilm interface. Water Res. 36, 4775–4784.

Watson S.W., Bock E., Valois, F.W., Waterbury, J.B., Schlosser, U., 1986. Nitrospira marina gen. nov., sp. nov.,: a chemolithotrophic nitriteoxidizing bacterium. Arch. Microbiol. 144, 1–7.

Windey, K., Bo, I.D., Verstraete, W., 2005. Oxygen-limited autotrophic nitrification–denitrification (OLAND) in a rotating biological contactor treating high-salinity wastewater. Water Res. 39, 4512–4520.

Ye, L., Tang, B., Zhao, K.F., Pijuan, M., Peng, Y.Z., 2009. Nitrogen removal via nitrite in domestic wastewater treatment using combined salt inhibition and on-line process control. Water Sci. Technol. 60(6), 1633–1639.

Zhu, S., Chen, S., 2001. Impacts of reynolds number on nitrification biofilm kinetics. Aquac. Eng. 24, 213–229.

Thesis defence presentation

Karlsruhe Institute of Technology

NITRIFIKATION VON SALZHALTIGEM ABWASSER IN FESTBETTREAKTOREN

SUDARNO

Institut für Ingenieurbiologie und Biotechnologie des Abwassers
Fakultät für Bauingenieur-, Geo- und Umweltwissenschaften

06. Mai 2011

Gliederung

Karlsruhe Institute of Technology

1. Einleitung
   - Hintergrund
   - Zielsetzung
2. Material und Methoden
   - Reaktoren für die Anreicherung von Nitrifikanten
   - Festbettreaktoren für die Biofilmbildung von halophilen Nitrifikanten
   - Festbettreaktoren mit wechselndem Salzgehalt
   - Kleine Reaktoren für Batch-Ansätze
3. Ergebnisse
   - Der Effekt von wechselnden Salzgehalten auf die Nitrifikation
   - Der Effekt von Temperatur, Ammonium und Nitrit – Konzentration auf die halophile Nitrifikation
4. Zusammenfassung

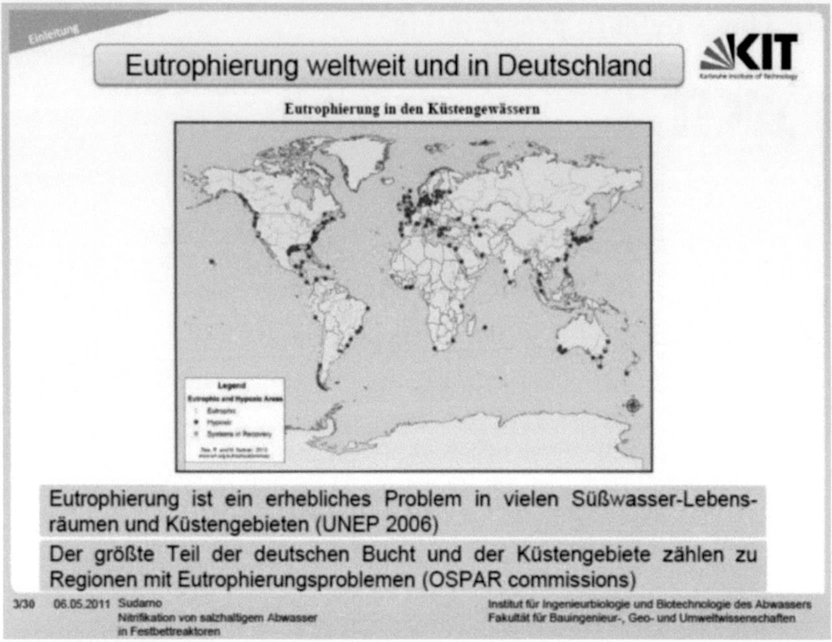
Einleitung
Eutrophierung weltweit und in Deutschland
KIT
Eutrophierung in den Küstengewässern
Legend
Eutrophierung ist ein erhebliches Problem in vielen Süßwasser-Lebensräumen und Küstengebieten (UNEP 2006)
Der größte Teil der deutschen Bucht und der Küstengebiete zählen zu Regionen mit Eutrophierungsproblemen (OSPAR commissions)
3/30 06.05.2011 Sudarno
Nitrifikation von salzhaltigem Abwasser
in Festbettreaktoren
Institut für Ingenieurbiologie und Biotechnologie des Abwassers
Fakultät für Bauingenieur-, Geo- und Umweltwissenschaften

Einleitung
Beitrag der verschiedenen anthropogenen Quellen zur Freisetzung von Stickstoff in Deutschland
KIT
Karlsruhe Institute of Technology
Haushalte ohne Abwasseranschluss 0,98%
Aquakultur 0,02%
Industrie 2,43%
Diffuse Einträge 10,69%
Landwirtschaft 58,64%
Kläranlagen 27,24%
5% weltweit unter halophilen Bedingungen
www.umweltbundesamt.de
Die Reduktion von Stickstoff aus Kläranlagen-Abwässern kann die Freisetzung von Stickstoff deutlich minimieren
4/30 06.05.2011 Sudarno
Nitrifikation von salzhaltigem Abwasser
in Festbettreaktoren
Institut für Ingenieurbiologie und Biotechnologie des Abwassers
Fakultät für Bauingenieur-, Geo- und Umweltwissenschaften

Einleitung

Biologische Stickstoffelimination aus Abwasser

KIT Karlsruhe Institute of Technology

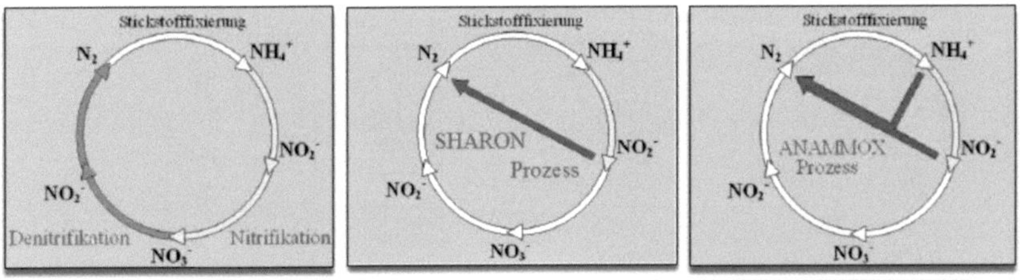

Die traditionellen N-eliminierenden Prozesse

Nitrifikation

i. $2NH_4^+ + 3O_2 \rightarrow 2NO_2^- + 4H^+ + 2H_2O$

(Ammonium oxidierende Bakterien – AOB)

ii. $2NO_2^- + O_2 \rightarrow 2NO_3^-$

(Nitrit oxidierende Bakterien – NOB)

Denitrifikation

i. $NO_3^- + 2H^+ \rightarrow NO_2^- + H_2O$

ii. $NO_2^- + 4H^+ \rightarrow \frac{1}{2} N_2 + 2H_2O$

SHARON Prozess

Nitritation

$2NH_4^+ + 3O_2 \rightarrow 2NO_2^- + 4H^+ + 2H_2O$

Denitritation

$NO_2^- + 4H^+ \rightarrow \frac{1}{2} N_2 + 2H_2O$

ANAMMOX Prozess

Nitritation

$2NH_4^+ + 3O_2 \rightarrow 2NO_2^- + 4H^+ + 2H_2O$

Anammox

$NH_4^+ + NO_2^- \rightarrow N_2 + 2H_2O$

Einleitung

Salzhaltige Abwässer

KIT Karlsruhe Institute of Technology

Fischverarbeitung

Lederherstellung

Erdölindustrie

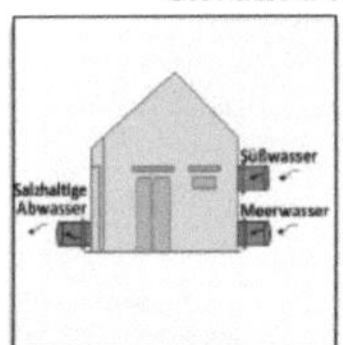

Duale Wasserversorgung

⇩ ⇩ ⇩ ⇩

Abwasser mit einer hohen Salz- und Ammoniumkonzentration

Tätigkeiten	Salzgehalt (%)	Ammonium (mg-N L^{-1})	Quelle
Fischverarbeitung	2,5 – 3	0,039 – 1940	Aspe et al. 1997
Lederherstellung	2,7	1200	Lefebvre et al. 2005
Erdölindustrie	5	80	Dahl et al. 1997
Duale Wasserversorgung	0,55	130	Tang et al. 2007
Kommunales Abwasser	0,01	40	Metcalf & Eddy 2003

Halotolerante oder halophile Nitrifikanten müssen vorhanden sein, um salzhaltige Abwässer zu nitrifizieren

Einleitung

Halophile und halotolerante Bakterien im salzhaltigen Abwasser

KIT Karlsruhe Institute of Technology

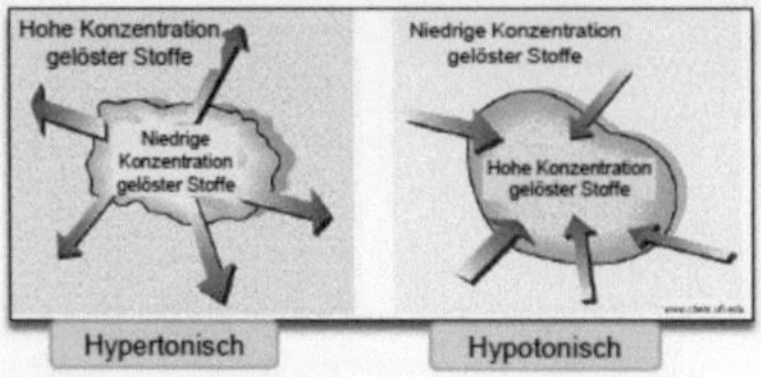

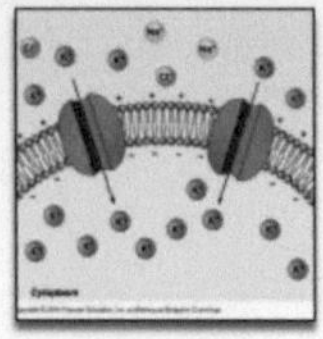

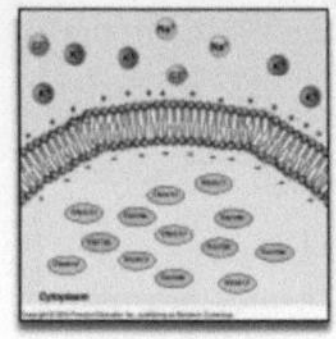

Osmotischer Druck:
- Hypertonisch → austrocknend
- Hypotonisch → anschwellend

Verringerung der Aktivität von Bakterien

Um den osmotischen Druck im Gleichgewicht zu halten:
- Kalium-Pumpe
- Synthese der kompatiblen Solute

Es wird zusätzliche Energie benötigt → daraus resultiert eine langsame Wachstumsrate

7/30 06.05.2011 Sudarno
Nitrifikation von salzhaltigem Abwasser in Festbettreaktoren
Institut für Ingenieurbiologie und Biotechnologie des Abwassers
Fakultät für Bauingenieur-, Geo- und Umweltwissenschaften

Einleitung

Zielsetzung

KIT Karlsruhe Institute of Technology

1. Das Finden von geeigneten Inokula für die Nitrifikation von salzhaltigen Abwässern
2. Vergleich des Einflusses unterschiedlicher Salzkonzentrationen auf Ammonium und Nitrit-Oxidationsraten (AOR und NOR) in Festbettreaktoren, die mit einer Mikroflora aus entweder Süßwasser oder Meerwasser beimpft wurden
3. Bestimmung der AOR und NOR von halophilen Nitrifikanten in Biofilmen bei verschiedenen Temperaturen
4. Auswirkung von verschiedenen Ammonium - und Nitrit Konzentrationen auf die AOR und NOR

8/30 06.05.2011 Sudarno
Nitrifikation von salzhaltigem Abwasser in Festbettreaktoren
Institut für Ingenieurbiologie und Biotechnologie des Abwassers
Fakultät für Bauingenieur-, Geo- und Umweltwissenschaften

Material & Methoden

Batch-Ansatz für die Anreicherung von Nitrifikanten

KIT Karlsruhe Institute of Technology

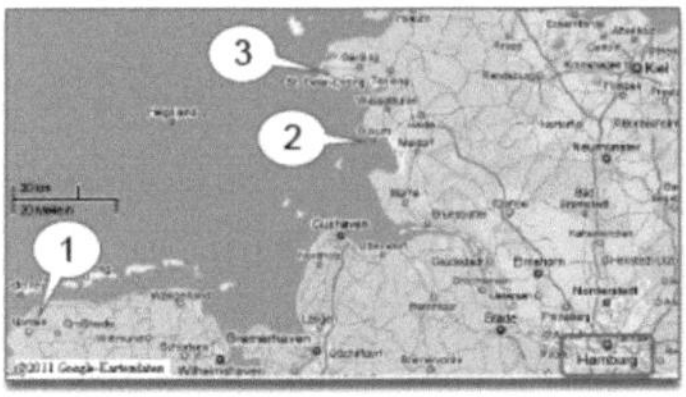

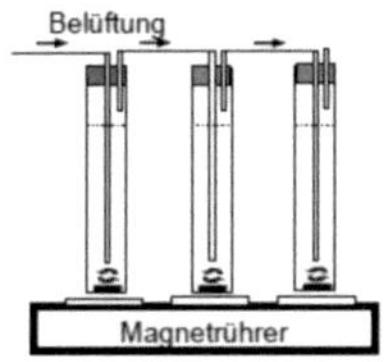

Probe	Salzgehalt (%)	Alkalinität (mg L^{-1}) als $CaCO_3$	NH_4^+-N (mg L^{-1})	NO_2^--N (mg L^{-1})	NO_3^--N (mg L^{-1})	Anorganische N (mg L^{-1})
1	3,75	400	0,2	0,32	10,3	10,83
2	3,19	120	0	0,11	2,3	2,43
3	3,21	130	0,6	0,68	1,7	2,98

- Drei Meerwasser & Schlamm-Proben aus Norden (Probe 1), vom Hafen in Büsum (2) und aus St. Peter–Ording (3) wurden verwendet
- Die autochthonen Mikroorganismen wurden mit 50 mg NH_4^+-N L^{-1} aktiviert
- Arbeitsvolumen: 250 ml; DO > 5 mg L^{-1}; 60 Tage

9/30 06.05.2011 Sudarno
Nitrifikation von salzhaltigem Abwasser in Festbettreaktoren
Institut für Ingenieurbiologie und Biotechnologie des Abwassers
Fakultät für Bauingenieur-, Geo- und Umweltwissenschaften

Ergebnisse

Anreicherung von Nitrifikanten

KIT Karlsruhe Institute of Technology

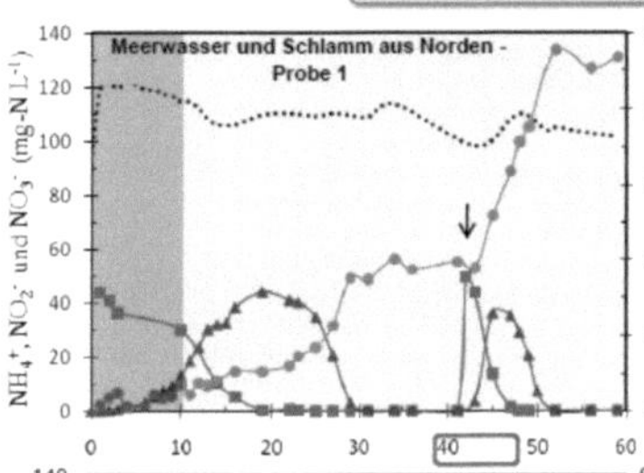

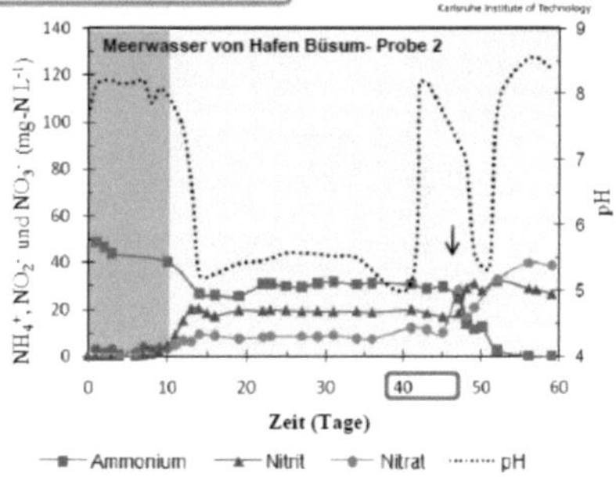

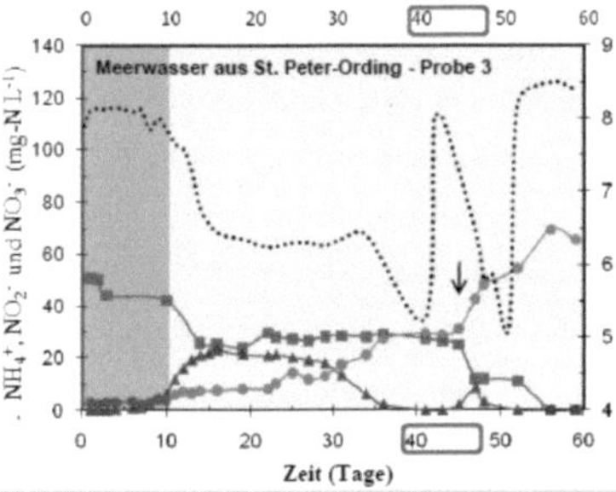

- Die Nitrifikation begann nach einer Lag Phase von 10 Tagen und benötigt einen pH-Wert über 7,0
- Die Ammoniumoxidation zu Nitrit war schneller im Vergleich zur Nitritoxidation zu Nitrat. Nach pH Korrektur wurde das gesamte Nitrit zu Nitrat oxidiert
- Umsatzraten der Probe 1, 2 und 3 betragen jeweils 15 mg NH_4^+-N L^{-1} $Tage^{-1}$, 11,5 und 6,6

10/30 06.05.2011 Sudarno
Nitrifikation von salzhaltigem Abwasser in Festbettreaktoren
Institut für Ingenieurbiologie und Biotechnologie des Abwassers
Fakultät für Bauingenieur-, Geo- und Umweltwissenschaften

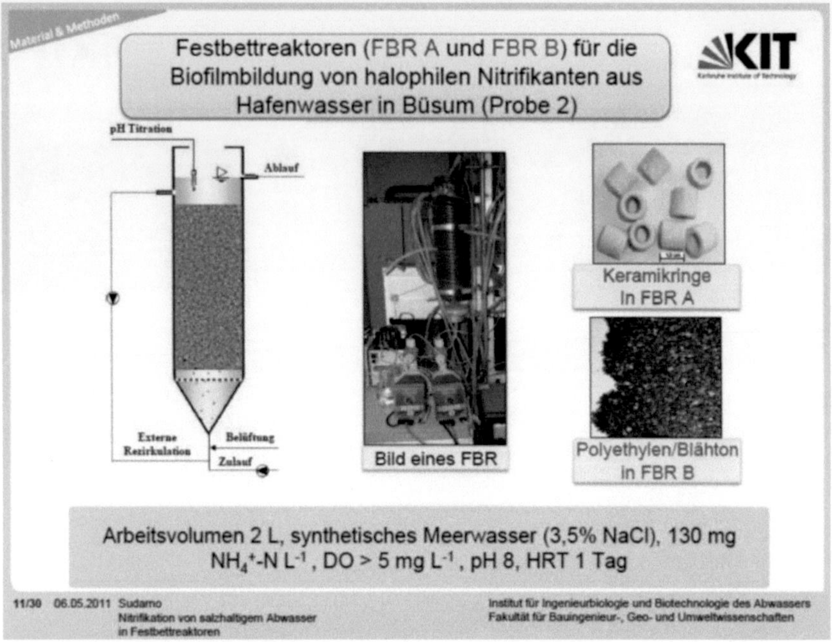
Material & Methoden
Festbettreaktoren (FBR A und FBR B) für die Biofilmbildung von halophilen Nitrifikanten aus Hafenwasser in Büsum (Probe 2)
KIT
pH Titration
Ablauf
Externe Rezirkulation
Belüftung
Zulauf
Bild eines FBR
Keramikringe in FBR A
Polyethylen/Blähton in FBR B
Arbeitsvolumen 2 L, synthetisches Meerwasser (3,5% NaCl), 130 mg NH_4^+-N L^{-1}, DO > 5 mg L^{-1}, pH 8, HRT 1 Tag
11/30 06.05.2011 Sudarno
Nitrifikation von salzhaltigem Abwasser in Festbettreaktoren
Institut für Ingenieurbiologie und Biotechnologie des Abwassers
Fakultät für Bauingenieur-, Geo- und Umweltwissenschaften

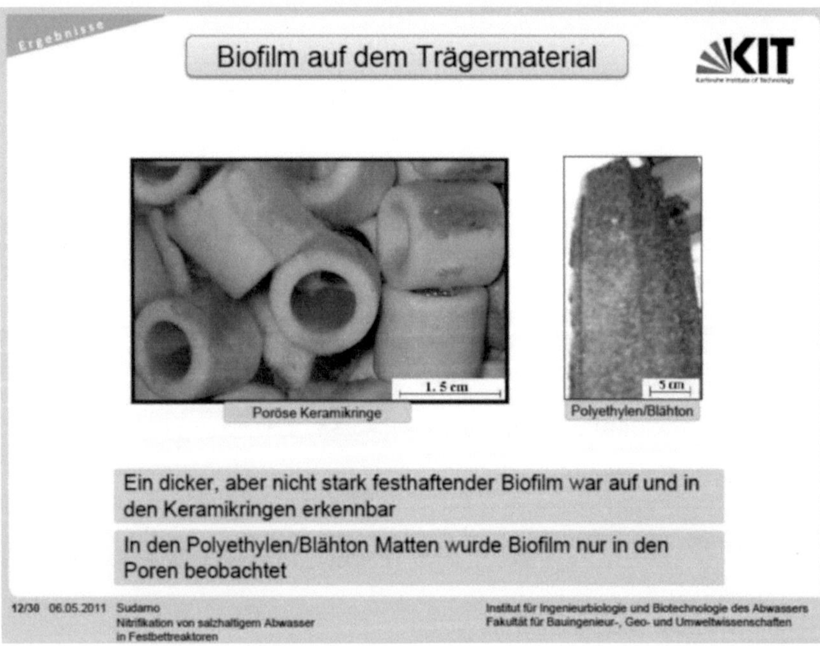
Ergebnisse
Biofilm auf dem Trägermaterial
KIT
1. 5 cm
Poröse Keramikringe
5 cm
Polyethylen/Blähton
Ein dicker, aber nicht stark festhaftender Biofilm war auf und in den Keramikringen erkennbar
In den Polyethylen/Blähton Matten wurde Biofilm nur in den Poren beobachtet
12/30 06.05.2011 Sudarno
Nitrifikation von salzhaltigem Abwasser in Festbettreaktoren
Institut für Ingenieurbiologie und Biotechnologie des Abwassers
Fakultät für Bauingenieur-, Geo- und Umweltwissenschaften

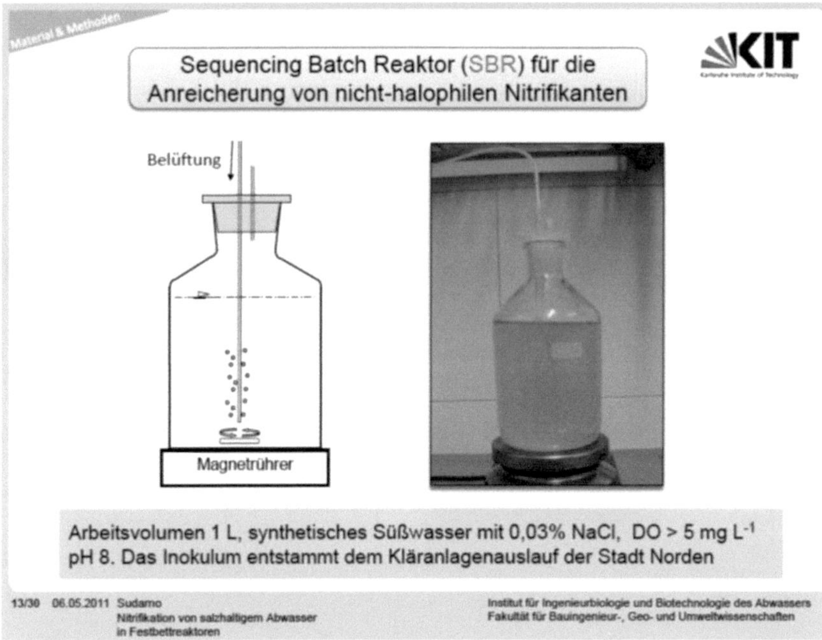
Material & Methoden
Sequencing Batch Reaktor (SBR) für die Anreicherung von nicht-halophilen Nitrifikanten
KIT
Belüftung
Magnetrührer
Arbeitsvolumen 1 L, synthetisches Süßwasser mit 0,03% NaCl, DO > 5 mg L-1
pH 8. Das Inokulum entstammt dem Kläranlagenauslauf der Stadt Norden
13/30 06.05.2011 Sudarno
Nitrifikation von salzhaltigem Abwasser in Festbettreaktoren
Institut für Ingenieurbiologie und Biotechnologie des Abwassers
Fakultät für Bauingenieur-, Geo- und Umweltwissenschaften

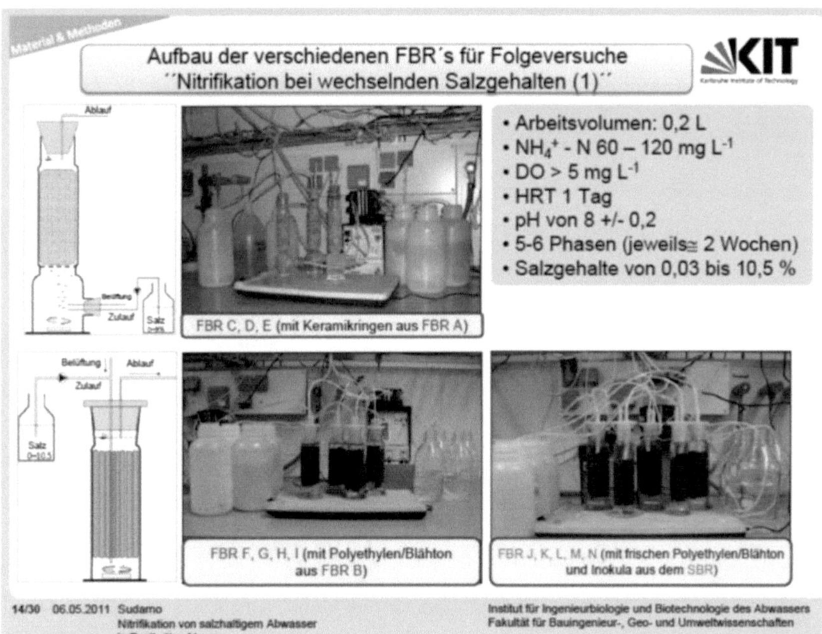
Material & Methoden
Aufbau der verschiedenen FBR´s für Folgeversuche
´´Nitrifikation bei wechselnden Salzgehalten (1)´´
KIT
Ablauf
Belüftung
Zulauf
Salz
• Arbeitsvolumen: 0,2 L
• NH4+ - N 60 – 120 mg L-1
• DO > 5 mg L-1
• HRT 1 Tag
• pH von 8 +/- 0,2
• 5-6 Phasen (jeweils≅ 2 Wochen)
• Salzgehalte von 0,03 bis 10,5 %
FBR C, D, E (mit Keramikringen aus FBR A)
FBR F, G, H, I (mit Polyethylen/Blähton aus FBR B)
FBR J, K, L, M, N (mit frischen Polyethylen/Blähton und Inokula aus dem SBR)
14/30 06.05.2011 Sudarno
Nitrifikation von salzhaltigem Abwasser in Festbettreaktoren
Institut für Ingenieurbiologie und Biotechnologie des Abwassers
Fakultät für Bauingenieur-, Geo- und Umweltwissenschaften

Material & Methoden

Nitrifikation bei wechselnden Salzgehalten (2)

KIT

FBR A (halophil) mit Keramikringen	FBR B (halophil) mit Polyethylen / Blähton	SBR (nicht halophil) mit Polyethylen / Blähton
FBR C, D, E (Beginn bei 3,5% NaCl)	FBR F, G, H, I (Beginn bei 3,5% NaCl)	FBR J, K, L, M, N (Beginn bei 0,03% NaCl)

FBR	Phase - NaCl (%)					
	I	II	III	IV	V	VI
C & F	3,5	2	0,03	3,5	5	3,5
D & G	3,5	1	0,03	3,5	7	3,5
E & H	3,5	0,5	0,03	3,5	9	3,5
I*	3,5	3,5	3,5	3,5	3,5	3,5

FBR	Phase - NaCl (%)				
	I	II	III	IV	V
J	0,03	0,5	0,03	10,5	0,03
K	0,03	1	0,03	9	0,03
L	0,03	2	0,03	7	0,03
M	0,03	3,5	0,03	5	0,03
N*	0,03	0,03	0,03	0,03	0,03

* : Kontrolle

15/30 06.05.2011 Sudarno
Nitrifikation von salzhaltigem Abwasser in Festbettreaktoren

Institut für Ingenieurbiologie und Biotechnologie des Abwassers
Fakultät für Bauingenieur-, Geo- und Umweltwissenschaften

Ergebnisse

FBR C, D und E mit Keramikringen, Beginn mit 3,5% NaCl

KIT

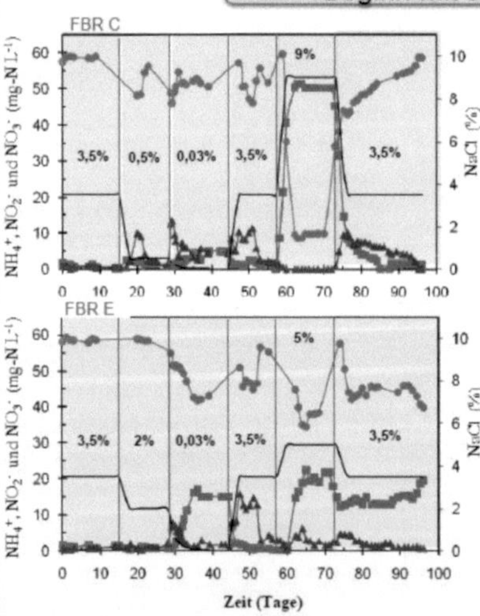

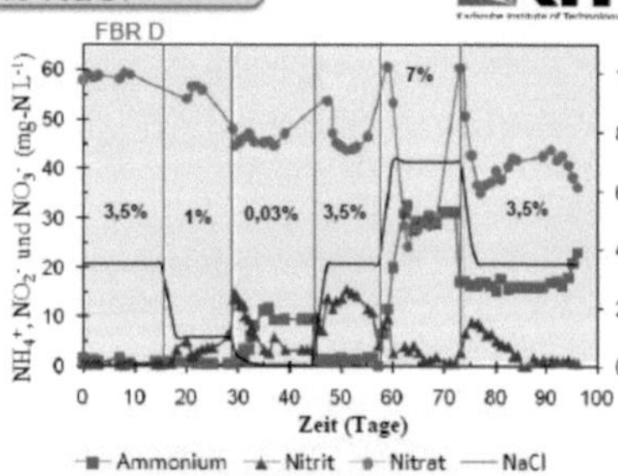

- Die NOB reagieren empfindlicher als die AOB
- Nach einigen Tagen der Anpassung des Biofilms an den neuen Salzgehalt wurde Ammonium vollständig zu Nitrat oxidiert
- Je höher der Salzgehalt ist, desto stärker ist die Hemmung auf die AOB
- Die Hemmwirkung auf die Ammoniumoxidation ist unter hypertonischen Bedingungen stärker als unter hypotonischen Bedingungen

16/30 06.05.2011 Sudarno
Nitrifikation von salzhaltigem Abwasser in Festbettreaktoren

Institut für Ingenieurbiologie und Biotechnologie des Abwassers
Fakultät für Bauingenieur-, Geo- und Umweltwissenschaften

Ergebnisse

Ammonium- und Nitritabnahme in FBR C, D und E mit maximaler NaCl-Konzentration von 9, 7 und 5 %

KIT Karlsruhe Institute of Technology

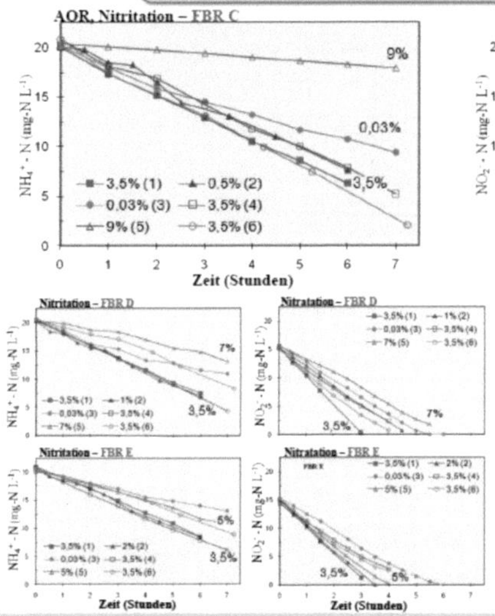

Die Abnahme der Ammonium - oder Nitrit-Konzentration bei dem jeweiligen Salzgehalt verlief linear in Laufe der Versuche

Die AOR und NOR wurden über die jeweilige Steigung der linearen Funktion bestimmt

17/30 06.05.2011 Sudarno
Nitrifikation von salzhaltigem Abwasser in Festbettreaktoren

Institut für Ingenieurbiologie und Biotechnologie des Abwassers
Fakultät für Bauingenieur-, Geo- und Umweltwissenschaften

Ergebnisse

AOR und NOR in FBR C, D und E

KIT Karlsruhe Institute of Technology

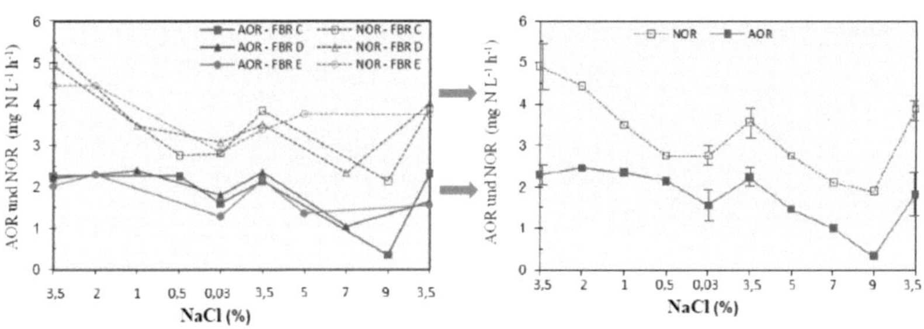

Die NOR war immer höher als die AOR

Bei abnehmendem Salzgehalt verringerte sich die NOR mehr als die AOR
Die Abnahme der AOR und der NOR war bei zunehmendem Salzgehalt fast gleich

Die AOB und NOB besitzen die Fähigkeit, sich von Salz-Stress zu erholen

18/30 06.05.2011 Sudarno
Nitrifikation von salzhaltigem Abwasser in Festbettreaktoren

Institut für Ingenieurbiologie und Biotechnologie des Abwassers
Fakultät für Bauingenieur-, Geo- und Umweltwissenschaften

Ergebnisse

AOR & NOR in FBR F,G,H (Polyethylen/Blähton, Beginn bei 3.5% NaCl)

KIT Karlsruhe Institute of Technology

Zum Vergleich:

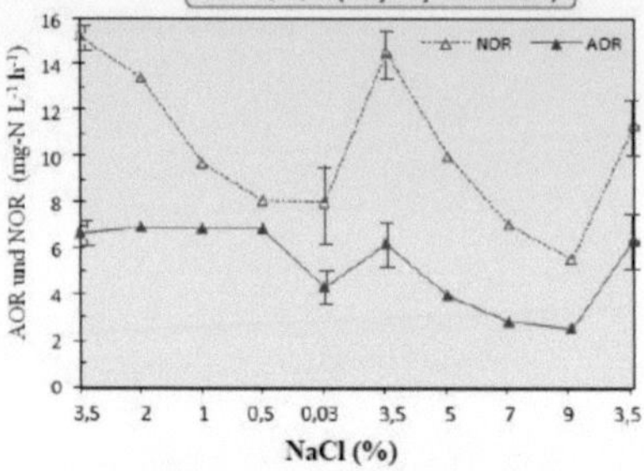

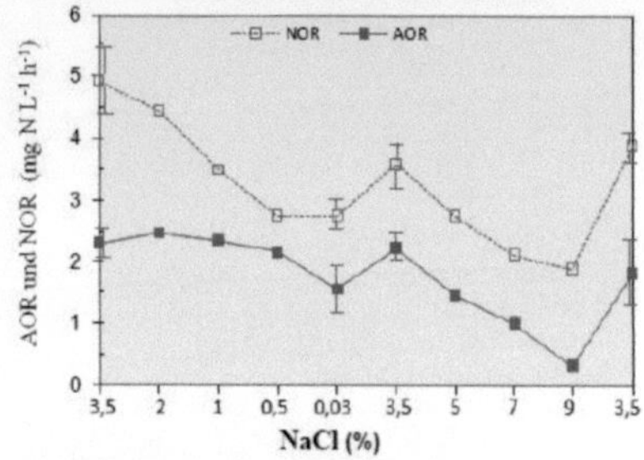

Der nitrifizierende Biofilm in den FBR´s mit Polyethylen/Blähton und -Keramikringen reagierten fast gleich auf den wechselnden Salzgehalt

Die Ammonium – und Nitritoxidationraten waren für die Bakterien in den FBR´s mit Polyethylen/Blähton etwa dreimal höher

19/30 06.05.2011 Sudarno
Nitrifikation von salzhaltigem Abwasser in Festbettreaktoren

Institut für Ingenieurbiologie und Biotechnologie des Abwassers
Fakultät für Bauingenieur-, Geo- und Umweltwissenschaften

Ergebnisse

FBR J, K, L und M - Polyethylen/Blähton mit nicht-halophilen Inokula, Beginn mit 0,03% NaCl

KIT Karlsruhe Institute of Technology

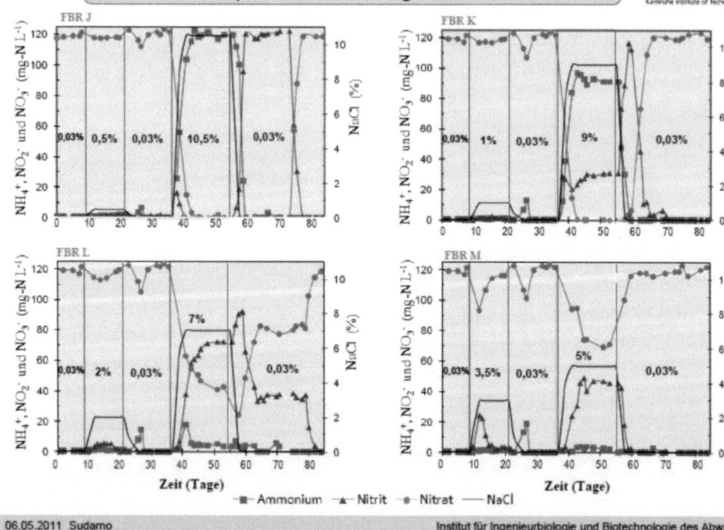

20/30 06.05.2011 Sudarno
Nitrifikation von salzhaltigem Abwasser in Festbettreaktoren

Institut für Ingenieurbiologie und Biotechnologie des Abwassers
Fakultät für Bauingenieur-, Geo- und Umweltwissenschaften

Ergebnisse

AOR und NOR in FBR J, K, L und M mit Polyethylen/Blähton mit nicht-halophilen Inokula, Beginn mit 0,03% NaCl

KIT Karlsruhe Institute of Technology

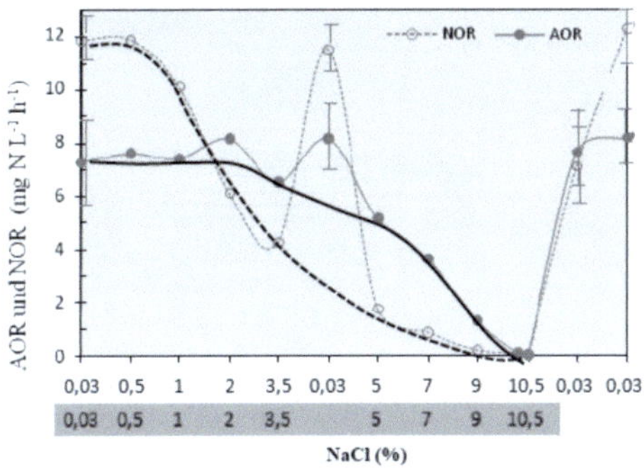

Ab 2 % NaCl war die NOR niedriger als die AOR und ab 9% NaCl wurden die NOB komplett gehemmt

Mit ansteigender NaCl-Konzentration war die NOR empfindlicher als die AOR

Bei hohem Salzgehalt war eine vorübergehende Hemmung in FBR´s, die mit nicht-halophilen Inokula beimpft wurden erkennbar

21/30 06.05.2011 Sudarno
Nitrifikation von salzhaltigem Abwasser in Festbettreaktoren

Institut für Ingenieurbiologie und Biotechnologie des Abwassers
Fakultät für Bauingenieur-, Geo- und Umweltwissenschaften

Ergebnisse

Vergleich der AOR und NOR in FBR´s mit nicht halophilen und halophilen Inokula

KIT Karlsruhe Institute of Technology

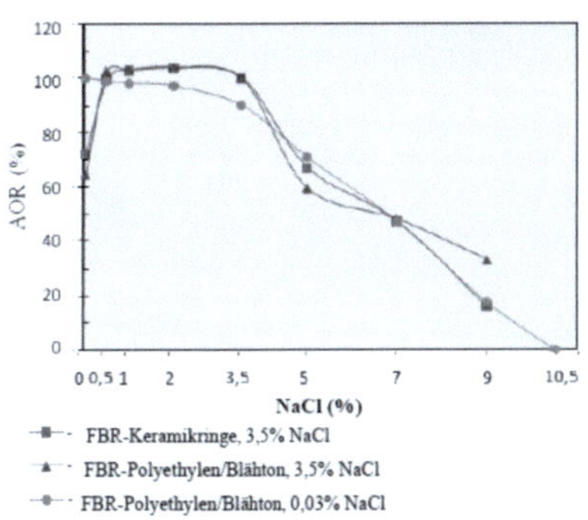

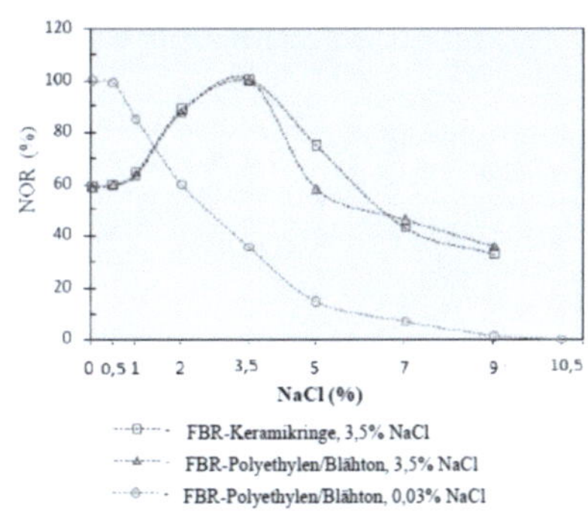

Die Aktivität von nicht halophilen und halophilen AOB ist ähnlich im Salz-Bereich von 0,5% bis 9%

Die halophilen NOB können auch bei höheren Salzkonzentrationen Nitrit mit hohen Raten oxidieren

22/30 06.05.2011 Sudarno
Nitrifikation von salzhaltigem Abwasser in Festbettreaktoren

Institut für Ingenieurbiologie und Biotechnologie des Abwassers
Fakultät für Bauingenieur-, Geo- und Umweltwissenschaften

Ergebnisse

Vergleich mit anderen Ergebnissen

Nitrifikanten	Reaktor	NaCl (%)	Prozentanteil der Rate (%)		Quelle
Angepaßter Belebtschlamm auf einen Salzgehalt von 3,3%	FR	3.3 bis 1.65 3.3 bis 5	140 (AOR) 70 (AOR)		Dahl et al. 1997
Angepaßter Belebtschlamm auf einen Salzgehalt von 0,5%	AS	0.5 bis 3 0.5 bis 7	70 (AOR) 20 (AOR)		Panswad und Anan 1999
Meerwasser	FBR	3.5 bis 1 3.5 bis 5 3.5 bis 7	103 (AOR); 75 (AOR); 60 (AOR);	78 (NOR) 65 (NOR) 46 (NOR)	Diese Arbeit
Nicht-adaptierter Belebtschlamm	AS	0 bis 3 0 bis 7	45 (AOR) 10 (AOR)		Panswad und Anan 1999
Nitrifizierender Belebtschlamm	SBR	0 bis 6.6 0 bis 1.65	0 45 (AOR);	 58 (NOR)	Moussa et al. 2006
Kommunales Abwasser	SBR	0 bis 2	25 (AOR);	0 (NOR)	Ye et al. 2009
Nitrosomonas & Nitrobacter	AS	0 bis 5	75 (AOR)		Dincer und Kargi 2001
Süßwasser	FBR	0 bis 3.5 0 bis 7	90 (AOR); 74 (AOR);	35 (NOR) 16 (NOR)	Diese Arbeit

FR = "Fluidized" Reaktor; FBR = Festbettreaktor;
AS = Belebtschlamm; SBR = Sequencing Batch Reaktor

Material & Methoden

Batch Ansätze zur Bestimmung des Einflußes von Temperatur, Ammonium und Nitrit auf die Ammonium- und Nitritoxidationsrate (AOR / NOR)

Keramikringe aus FBR A

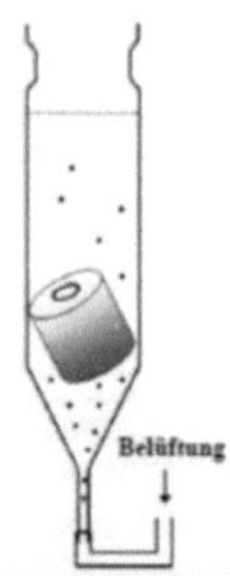

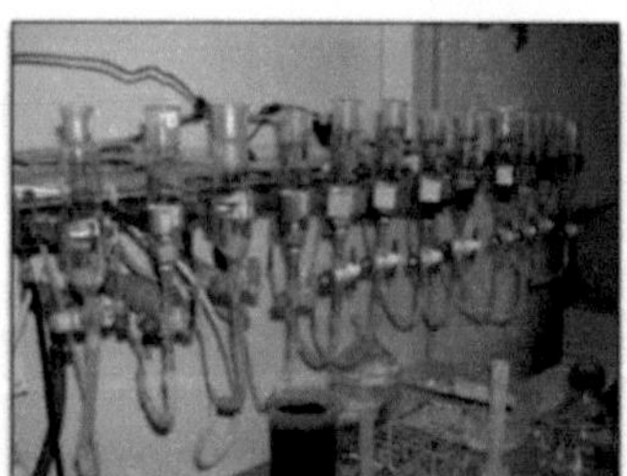

Messungen	Bereiche der getesteten Faktoren		
	Temperatur °C	Ammonium g N L⁻¹	Nitrit g N L⁻¹
AOR	6 – 50	0 – 5	0 – 0,4
NOR	6 – 50	0 – 5	0 – 5

Ergebnisse

Die Wirkung der Temperatur auf die AOR und die NOR

KIT Karlsruhe Institute of Technology

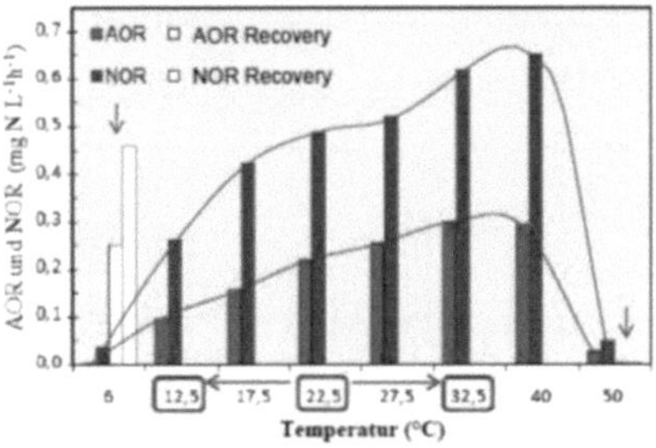

Ein Temperaturabfall von 22,5 auf 12,5 °C reduziert die AOR auf 44% und die NOR auf 55% der Anfangsrate (halbiert)

Ein Temperaturanstieg von 22,5 auf 32,5 °C erhöht die AOR auf 136% und die NOR auf 126% der Anfangsrate (um ~30 %)

Nitrifikanten	Temperatur-Bereich (°C)	Temperatur-Korrekturfaktor pro °C		Quelle
		AOR	NOR	
als Biofilm	8–27	1,043	-	Zhu and Chen 2002
in Suspension	10–23	1,098	-	Salvetti et al. 2006
in Suspension	10–20 10–20	1,027 bis 1,127 1,088 bis 1,116	-	Hwang und Oleszkiewicz 2007
in Suspension	5–20 20–35	1,172 1,062	-	Guo et al. 2010
als Biofilm	12,5–22,5	1,059	1,047	Diese Arbeit
als Biofilm	22,5–32,5	1,036	1,026	Diese Arbeit

Ergebnisse

Die Wirkung von Ammonium und Nitrit auf die AOR und die NOR

KIT Karlsruhe Institute of Technology

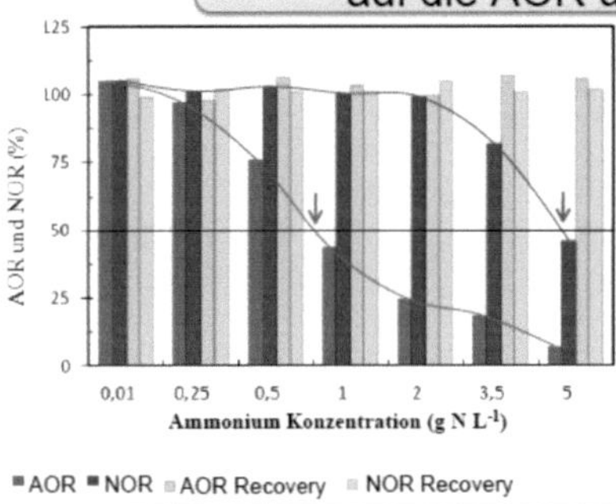

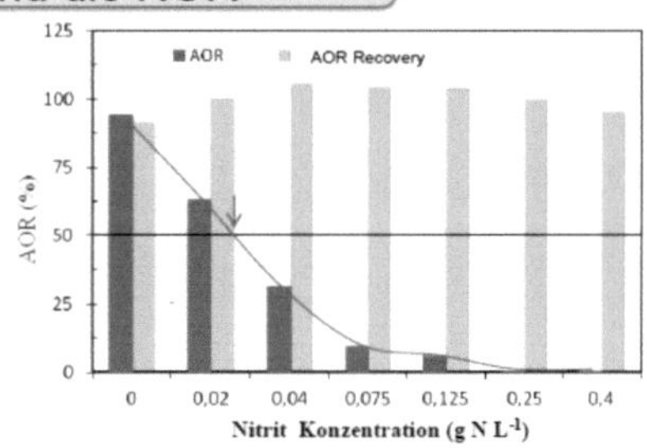

Ammonium hemmte die AOB stärker als die NOB

IC_{50} der AOR : 0,8 g NH_4^+-N L^{-1}
IC_{50} der NOR : 4,9 g NH_4^+-N L^{-1}

Ammonium bewirkte nur eine vorübergehende Hemmung der AOR und NOR

Keine Hemmung der NOR durch Nitrit, aber Hemmung der AOR
IC_{50} der AOR : 0,027 g NO_2^--N L^{-1}

Nitrit hatte bis zu einer Konzentration von 0,4 g NO_2^--N L^{-1} nur eine vorübergehende Hemmwirkung auf die NOR

IC_{50} : die Konzentration eines Inhibitors, bei der eine halbmaximale Inhibition beobachtet wird

Ergebnisse

FA oder FNA Anfangskonzentration, ab der eine Hemmung der AOB oder NOB stattfindet (Literaturvergleich)

KIT Karlsruhe Institute of Technology

$$FA\left(NH_3\, mg\, L^{-1}\right) = \frac{17}{14} \times \frac{[NH_4^+] \times 10^{pH}}{e^{\left[\frac{6344}{273+t}\right]} + 10^{pH}}$$

$$FNA\left(HNO_2\, mg\, L^{-1}\right) = \frac{46}{14} \times \frac{[NO_2^-]}{e^{\left[\frac{-2300}{273+t}\right]} \times 10^{pH}}$$

	FA mg N L^{-1}	FNA mg N L^{-1}	Bedingungen			Quelle
			pH	Temperatur (°C)	NaCl	
AOB	10 – 150		n.a	10 – 23	Niedrig	Anthonisen et al. 1976
	> 10		6,8 – 7,5	35	Niedrig	Mosquera-Corral et al. 2005
	1,4		8	22,5	3,5%	Diese Arbeit
		0.1	7,1	30	Niedrig	Vadivelu et al. 2006
		0.5	6,5	21	Niedrig	Qiao et al. 2010
		< 0,0015	8	22,5	3,5%	Diese Arbeit
NOB	0.1–10		n.a	10 – 23	Niedrig	Anthonisen et al. 1976
	>1		6,8–7,5	35	Niedrig	Mosquera-Corral et al. 2005
	110-190		8	22,5	3,5%	Diese Arbeit
		0.22–2,8	n.a	10-23	Niedrig	Anthonisen et al. 1976
		>> 0,79***	8	22,5	3,5%	Diese Arbeit

*** Keine Hemmung

27/30 06.05.2011 Sudarno
Nitrifikation von salzhaltigem Abwasser in Festbettreaktoren
Institut für Ingenieurbiologie und Biotechnologie des Abwassers
Fakultät für Bauingenieur-, Geo- und Umweltwissenschaften

Zusammenfassung

Zusammenfassung

KIT Karlsruhe Institute of Technology

1. Nitrifikanten aus dem Hafenwasser von Büsum können als Biofilm auf Keramikringen und Polyethylen/Blähton wachsen

2. Der wechselnde Salzgehalt hat eine stärkere Hemmung auf die NOB als auf die AOB bei kontinuierlichen- und Batch-Ansätzen in FBR´s mit halophilen und nicht-halophilen Inokula

3. Mit nicht halophilen Inokula nimmt die NOR bei steigenden Salzgehalten auf Werte unter der AOR ab und es reichert sich Nitrit an

4. In FBR´s mit halophilen Inokula wurde die Nitrifikation unter hypertonischen Bedingungen stärker gehemmt als unter hypotonischen Bedingungen

5. FBR´s mit halophilen Inokula können in einem breiteren Salz-Bereich (von 0,5% bis 3,5%) nitrifizieren als mit nicht-halophilen Inokula (von 0,03% bis 0,5%)

28/30 06.05.2011 Sudarno
Nitrifikation von salzhaltigem Abwasser in Festbettreaktoren
Institut für Ingenieurbiologie und Biotechnologie des Abwassers
Fakultät für Bauingenieur-, Geo- und Umweltwissenschaften

Zusammenfassung

Zusammenfassung

KIT Karlsruhe Institute of Technology

6. Die Hemmung der Aktivität von nicht-halophilen und halophilen AOB im Salz-Bereich von 3,5% bis 9% war ähnlich

7. Die halophilen NOB können Nitrit in einem breiteren Salz-Bereich als die nicht halophilen NOB eliminieren

8. Ein Temperaturanstieg von 22,5 °C auf 32,5 °C führte zu einer Verbesserung der AOR und NOR um einen Faktor von 1.036 und 1.026 pro Grad Celcius

9. AOB waren empfindlicher gegenüber steigenden Ammonium- und Nitrit-Konzentrationen als NOB

29/30 06.05.2011 Sudarno
Nitrifikation von salzhaltigem Abwasser
in Festbettreaktoren

Institut für Ingenieurbiologie und Biotechnologie des Abwassers
Fakultät für Bauingenieur-, Geo- und Umweltwissenschaften

Danksagungen

KIT Karlsruhe Institute of Technology

Diese Arbeit wurde durch ein Stipendium
des Deutschen Akademischen Austauschdienstes DAAD
und ein Stipendium der Max-Buchner Stiftung -
DECHEMA unterstützt

Vielen Dank für ihre Aufmerksamkeit

- Sudarno, U., Bathe, S., Winter, J., Gallert, C., 2010. Nitrification in fixed-bed reactors treating saline wastewater. Appl. Microbiol. Biotechnol. 85, 2017 – 2030
- Sudarno, U., Winter, J., Gallert, C., 2011. Effect of varying salinity, temperature, ammonia and nitrous acid concentrations on nitrification of saline wastewater in fixed-bed-reactors. Biores. Technol. 102, 5665 – 5673

30/30 06.05.2011 Sudarno
Nitrifikation von salzhaltigem Abwasser
in Festbettreaktoren

Institut für Ingenieurbiologie und Biotechnologie des Abwassers
Fakultät für Bauingenieur-, Geo- und Umweltwissenschaften